STRATEGIC SURVEY 1986-1987

THE INTERNATIONAL INSTITUTE FOR STRATEGIC STUDIES
23 Tavistock Street
London WC2E 7NQ
Telephone 01-379 7676

STRATEGIC SURVEY 1986–1987

Published by
The International Institute for Strategic Studies
23 Tavistock Street, London WC2E 7NQ

This publication has been prepared by the Director of the Institute and his Staff, who accept full responsibility for its contents. These do not, and indeed cannot, represent a consensus of views among the world-wide membership of the Institute as a whole.

First published Spring 1987

ISBN 0 86079 125 4
ISSN 0459-7230

Printed in Great Britain by Adlards & Son Ltd., The Garden City Press, Letchworth, Herts.
Bound by British Otabind Ltd, Colchester

CONTENTS

Perspectives

1986 was a year of sharp contrasts, particularly in the fortunes of the two major international powers: the United States and the Soviet Union. President Reagan began the year with the confidence that derived from his extraordinary popularity at home and substantial approval from his key allies abroad. He had enjoyed a considerable success at the Geneva summit at the end of 1985; he had persuaded the Congress to reverse its ban on military assistance to the Contras and thus assured, at least temporarily that this aspect of his foreign policy would move forward; he continued to enjoy the substantial support of the American people for his firm stand towards the Soviet Union and his quest for strategic defences; and he had placed the US in a position that made a significant arms-control agreement possible.

As the year wore on, however, his Administration began to look less in control of its policies, reacting to outside challenges with uncertainty and in the end failing to cope. Against the President's wishes, Congress imposed economic sanctions on South Africa. Then, manipulated by Gorbachev into the Reykjavik meeting and misjudging what the Soviet leader had planned, Reagan pursued his dreams about the Strategic Defense Initiative (SDI), rather than the more modest, but significant agreements in the talks on reductions in intermediate-range nuclear forces (INF) and the strategic arms reduction talks (START) that both sides had moved towards in the previous months. Finally, his Administration traduced its own policies on arms sales to Middle East belligerents and on hostages, reducing further than usual the effectiveness of any US Administration in its last two years of office. The sum of the year's activities, misadventures and revelations left the Administration with its foreign-policy goals undermined, and the President's own authority and credibility damaged.

By contrast, the Soviet Union's dynamic new leader, Mikhail Gorbachev, had some success in beginning to nudge Soviet policy out of the glacial inertia induced by his predecessors. While there might be considerable argument about the efficacy of his policies, or the strength of his personal power, there could be no question about the coherence with which he acted. He – and, by extension, the Soviet leadership – appeared to know where they were going and seemed to have a clear idea of how to get there.

As the year moved on, the pace of Gorbachev's initiatives increased. Three times he extended the unilateral Soviet nuclear test ban; he made proposals for major reductions in nuclear and conventional forces, not just in Europe but in Asia as well; and he showed at least rhetorical flexibility over on-site inspection and other verification provisions, thereby keeping the West on the defensive. He acknowledged the problems created by the continuing war in Afghanistan by first withdrawing a few Soviet troops from that country, then (through his new Afghan puppet, Dr Najibullah), declaring a six-month cease-fire and suggesting that a negotiated end to the war was close at hand. As part of a deal by which a Soviet spy was exchanged for a US newsman detained on trumped-up charges, he secured a meeting with the US President – not on American soil, as had been agreed at the Geneva meeting in 1985, and on his own terms.

Gorbachev's championing of *glasnost* or 'openness' in the Soviet Union made it appear that real reforms, rather than just adjustments, had been launched there. He broke all precedents with his personal call to Andrei Sakharov to inform him (and the world) that his exile in Gorky was at an end and that his welcome in Moscow would include the right to talk with the Western press.

The propaganda advantages which Gorbachev gained from such actions were manifest, but whether these will be translated into significant gains in the contest between the rival camps is more problematic. The non-Communist governments in both Europe and East Asia, which he has been wooing with increased ardour, still regard the Soviet Union with deep underlying suspicion. There has not yet been enough clear change within the Soviet Union to overcome the cynicism with which most people, including Soviet citizens, view the results of almost seventy years of Communist rule; reform will have to be unambiguous and sustained before it will be recognized. Gorbachev himself may be sincere in his efforts, but the continued strength of more conservative forces in the Soviet Union calls into question his ability to effect real change.

The outlook for the West in the near future remains unpromising. Gorbachev has demonstrated that he is a subtle and sophisticated politician, and he can be expected to gild many more Soviet policies with a new sheen that will challenge the wit and will of a weakened US government. President Reagan would have been in a better position to reassert his leadership if he could have been convinced at the very outset of the need to accept real responsibility for the mistaken Iran/Contra policy, to make a clean sweep of those who had advised him so poorly, and to replace them with men who could command the respect of country and Congress. But this is perhaps to demand a different type of leader: it was Reagan's approach and style of management that provided the soil in which the disarray could flourish. His failure to act has virtually ensured that the most powerful nation in the Western alliance will face the challenges of the coming year with its ability to conduct a forceful foreign policy badly compromised.

The Unravelling of Authority

Critics of President Reagan have often claimed to be puzzled by his immense popularity. They have pointed to his unwillingness or inability to master the details of complex international and domestic problems and to his constant misstatements in press conferences and *ad hoc* remarks that suggested an incomplete understanding of the issues. Putting him down as a second-rate actor, rather than the skilled politician he had become, they decried his adherence to a hard-line ideology, when it was often clear that his actions were as flexible as they needed to be to overcome obstacles as they appeared.

President Reagan had succeeded because he mirrored the aspirations of many of his countrymen. His high moral tone, his ingrained anti-Communism, his insistence that 'America must stand tall', matched the taste of much of the American electorate for straightforward solutions and strong action and therefore held a powerful appeal. A majority of the American public approved of his firm stance against the Communist threat, his efforts to build up US military strength, his decision to invade Grenada, the encouragement of democracy in Haiti and the Philippines, the reduction of the inflation rate and the economic recovery. In his symbolic role as leader of the nation, Ronald Reagan looked good, sounded confident and made many Americans feel proud. For this the public was prepared to overlook his imperfect grasp of the issues, his lax style of management and his important policy failures – including the accumulation of an enormous national debt, a disastrous trade deficit, and his ineffective handling of key foreign-policy challenges from Lebanon to Central America and South Africa. During 1986, however, the Administration stumbled. It found it more and more difficult to respond adequately to the growing complexities of foreign-policy problems. Faced with opposition from a more recalcitrant Congress, a more flexible opponent in Gorbachev, and a variety of intractable foreign problems, the Administration's lack of sophisticated ideas and processes began to take their toll.

Part of the problem lay in the structure of the Administration's policy-making apparatus, which proved woefully inadequate. Cabinet secretaries and departments sometimes pursued uncoordinated, if not contradictory, policies. The President's disregard for details and his refusal to give more than general policy guidance left much of the formulation and execution of policy to a small group of men close to him in the White House and the National Security Council. Successes early in the Administration's term bred overconfidence and a belief that the righteousness of its policies justified whatever means it saw fit to use. Above all, there was a conviction that the President's popularity was so great that he could always swing the country behind whatever the Administration did.

This confidence manifested itself in an unwillingness to compromise – most prominently over the case of economic sanctions against South Africa, but also over the level of defence expenditure, aid to the Nicaraguan Contras, and measures to increase taxes as a way to reduce the federal deficit. As a result, the Administration began to lose ground to its opponents on basic policy issues; perhaps more importantly, the Congress showed increasing critical interest in American defence and foreign policy as well as domestic economic policy. (However, although the Congress began to play, and will continue to play, an increasingly important role, it is neither constitutionally nor politically able to provide an overall policy framework and it cannot actually direct foreign policy).

What the President's men seem to have ignored is that support, even for a popular President, is grounded in a sense of success and respect for the law. Throughout 1986, though, problems were accumulating for the Administration's own foreign- and defence-policy agenda, and by the end of the year the Administration was clearly seen to be failing. When it identified Col. Gaddafi as one of the forces behind terrorism and sent military aircraft to bomb Libyan territory in retaliation, most of America applauded loudly, even if many allies and the rest of the world either stood mute or condemned the action. In this case, the Administration's policies achieved some success: the basis of Gaddafi's authority was threatened and he withdrew from the international stage, and America's allies took some additional steps to co-ordinate their policies and isolate Libya. But none of this solved the problems of terrorism or removed the threat of kidnappings, and the admission later in the year that the White House had been deeply engaged in a campaign of 'disinformation' against Gaddafi tarnished what had originally appeared to be a principled action.

A major goal of the Administration during the year had been to ensure that Gorbachev would fulfil his promise, given at the Geneva summit in 1985, to participate in another summit in the US in 1986. The two leaders, however, had somewhat different views on the content of such a summit: Reagan was apparently happy to see another friendly get-together, even if it reached no firm conclusion; Gorbachev, on the other hand, insisted on a clear agenda leading towards a comprehensive arms-control agreement, with particular emphasis on reining in Reagan's Strategic Defense Initiative. Gorbachev's reluctance to engage in another summit meeting, especially in the United States, without the promise of a new formal agreement was no doubt buttressed by the concerns of the more conservative forces in the Soviet Union, including the military.

In late August the KGB's arrest and threatened trial on fabricated charges of Moscow-based American reporter Nicholas Daniloff, in retaliation for the arrest of a low-level Soviet spy in the US, brought about a confrontation. Daniloff's incarceration placed the Administration in an unenviable dilemma. His arrest infuriated American public opinion and put enormous pressure on the President to secure his release, yet the Soviet price – exchanging Daniloff for a real spy –

was too high for Reagan to accept outright. In the event, Daniloff was freed as part of a complex deal, in which the Soviet spy was returned to the USSR one day later, and an agreement was reached that Reagan and Gorbachev would meet in Reykjavik. The US President continued to insist that he had not bartered the freedom of an innocent newsman for that of a spy, but many found it hard to believe that the *quid pro quo* was not exactly what the Soviet Union had demanded.

Nor did the putative summit meeting in Reykjavik the next month do much to restore Reagan's reputation. Believing it was going to be simply an intermediate stop on the road to a full-fledged summit in the US later in the year or in 1987, the President arrived in Reykjavik expecting to lay out his proposals on START and SDI, agree to guidelines for an 'interim' INF agreement, and set the date for a summit in the US. The Soviet leader, however, moved immediately to try to negotiate a wide-ranging and ambitious agreement on arms control. The USSR also exploited the meeting to the full for its public-relations effect. The Soviet party, unlike the American, arrived with a full panoply of public-relations and propaganda specialists, press secretaries and high-level analysts – all prepared, even eager, to meet the Western media and present the Soviet position in the best possible light. Although the White House subsequently made a strong effort to present the meeting as a major success for Reagan and the US, it appeared at the time that an opportunity to achieve a breakthrough in arms control had been lost.

The extent of the erosion of the Reagan magic could be judged by the Congressional elections in early November. Although the President threw himself wholeheartedly into the contest, and asked the people to show their support for him by supporting Republican candidates, the result was a significant loss of power. The Democrats regained control of the Senate by a margin considerably greater than even their own expectations. With both houses of Congress securely in Democrat hands, the last two years of Reagan's presidency were clearly going to be difficult for him and his party.

In November came revelations that the Administration had for some time been secretly supplying arms to Iran, while putting pressure on others not to do so; that behind the scenes it had been dealing with terrorists to try to get American hostages released, while publicly proclaiming its intention of never doing such a thing; and that it had been secretly (and perhaps illegally) supplying money and arms to the Nicaraguan Contras when the Congress had clearly forbidden such supply. This was the climax to a difficult year for Reagan. His credibility was severely tested, and it was soon clear that the majority felt that he had been lying to the country – something that Americans do not take lightly. More serious than the loss of an individual President's credibility, however, was the loss of the country's. Its ability to affect events in the Middle East, with regard to terrorism, the Iran–Iraq war and the Arab–Israeli conflict, not to mention Central America, will not be easily recovered.

In the immediate aftermath of the Iran/Contra scandal the President seemed unwilling to recognize publicly the extent of the damage that had been done and unable to redefine his objectives so as to re-establish a public consensus on the objectives of American foreign policy. And at the very moment when he needed to have his government acting together, it fragmented. After the publication of the highly critical Tower report on 26 February 1987, the President did take greater responsibility in a useful address to the nation. And it helps that a new, and respected, White House Chief-of-Staff and Assistant for National Security Affairs have been appointed. But they cannot be expected to do the job as it needs to be done until the President is prepared not just to re-establish his authority but also to give direction in foreign policy. America's allies have good cause for their concern, uncertain as to what to expect and worried about what might happen if the West is seriously tested by the more dynamic Soviet leader during 1987.

The Struggle For Authority

While President Reagan was losing his grip on the levers of authority, General Secretary Gorbachev was trying to tighten his own. Since the early days of his accession to power, Gorbachev had made his priorities abundantly clear. A rejuvenation of political and economic life in the Soviet Union was needed if the country was not to sink further into the morass of stagnation which was the legacy of the Brezhnev, Andropov and Chernenko regimes. For this, a certain stability in world affairs was called for; there need not be a return to detente, but an emphasis on the necessity for arms control became the core element in Soviet policy towards the West.

If wide-ranging arms-control agreements could be reached, Gorbachev argued, particularly with regard to the SDI, the need for heavy new investments in offensive and defensive nuclear arms would be obviated, making it possible to improve the civilian economy. Foreign-policy goals included a reassertion of control in Eastern Europe and efforts to divide the United States from its allies in Europe and East Asia. The Third World was lowest in the scale of priorities: new initiatives would be confined to attempts to consolidate the Soviet positions in Afghanistan, India and the Arab countries, and the emphasis would be on maintaining old gains, rather than reaching out for new and perhaps risky ones.

What Gorbachev showed above all was an acute grasp of the grave difficulties facing Soviet society. His announced goals have an inner logic that would appeal to most observers, and if they could be achieved they would be welcomed by most Soviet citizens. Yet their attainment demands fundamental changes in the way the Soviet Union is governed – and thus the wholehearted co-operation of his colleagues in the Politburo and subordinates throughout the bureaucratic structure. Gorbachev's own actions over the past eighteen months have strongly suggested that these conditions have yet to be fulfilled.

If bedrock reforms have not been achieved, or even attempted, there has nonetheless been a remarkable change in the tone of Soviet life. The tight constraints that have always bound the Soviet media have been loosened, and it has been encouraged (within careful limits) to report shabby practices, corruption and high-level peculation. Speeches by Gorbachev and his closest allies have sharply criticized past practices, with particular emphasis on Brezhnev's period of rule. This serves two linked purposes: it points the direction that Gorbachev sees as essential for correcting work habits and restructuring the economy, and it calls into question the past conduct of the remaining members of the Politburo, appointed by Brezhnev, who continue to obstruct Gorbachev's efforts.

In many respects, the explosion at the Chernobyl nuclear power plant illustrated how difficult it is to overcome the obduracy of traditional Soviet bureaucracy. Although Gorbachev had just announced the need for 'openness' in reporting failures, there was no official Soviet admission for three days that anything had even gone wrong, and Gorbachev himself did not address the nation for almost three weeks. It is probably true that the facts were initially kept even from the Politburo, but that could not have been the case for long. To be sure, when the system finally recovered its poise, the flow of information and criticism was unprecedented. But that it took so long to appear indicated that, over and above the predictable stonewalling of those involved at the local level, there must have been disagreement at the very top on how to handle the issue.

That Gorbachev also twice postponed the Central Committee meeting which should have been held in October 1986 was another indication of the difficulties he faces. When it finally met, at the end of January 1987, the Committee advanced Gorbachev's public campaign further than most analysts of Soviet affairs had expected. His call for greater democracy, including some elections in the lower levels of the party structure which would offer a choice of candidates, made it clear that he had succeeded in gaining a measure of agreement from the remaining hard-liners in the Politburo for different norms of Party practice. Reforms such as these are necessary if the Soviet people are to make their contribution to development willingly. All the same, it is too early to tell for certain whether the change in language will be matched by a significant change in practice. Gorbachev still faces an uphill struggle, but he may now feel more confident that the gradient is becoming less steep.

For the West the changes that Gorbachev is effecting offer both opportunity and challenge. He has already shown that he is a more subtle and dynamic opponent than Western leaders have had to face for many years. If his programmes for the rejuvenation of Soviet life are successful, the USSR will grow in strength and confidence, and, unless there is a change in the Soviet view of the nature of the international competition, in the long run the West will face a much stronger rival. Yet if the Soviet Union does become a more open society, if it finds it best to devote its resources more to domestic economic and political change, if it recognizes that it has much to gain from interdependence in the world economic system, then internal reforms may be followed by changes in Soviet behaviour abroad. The West would be well advised to view what is developing with a very open mind, cautious when meeting Soviet initiatives, but ready to welcome and encourage the possibility that a more normal relationship might replace that of outright antagonism.

A Lost Opportunity?

During the past year, the essence of super-power relations was arms control. President Reagan's efforts to expand the agenda to include regional and human-rights issues were unavailing. For obvious reasons the Soviet Union rejected his approach, and even Western public opinion saw more virtue in efforts to bring the super-power military competition under control than in attempts to improve human rights in Eastern Europe or to halt regional conflicts.

That arms control played a central role in East–West relations was not unusual. What was unusual, though, was that the customary linkage between arms control and other tensions between the two powers was reversed. In previous years, if problems arose in other areas, then efforts to negotiate agreements on arms suffered as a result: the Soviet invasion of Afghanistan and the shooting-down of the Korean airliner had an immediate effect on arms-control negotiations. But in 1986 Gorbachev, by extending his unilateral moratorium on nuclear testing, used arms control to deflect attention from his handling of Chernobyl. Reagan used arms control, and the announcement of the meeting in Reykjavik, to deflect attention from his handling of the Daniloff affair. It cannot be said that either effort was wholly successful, but at least arms-control negotiations in 1986 were not, as in years past, held hostage to other difficulties in Soviet–American relations.

The inability of the two powers to reach even minimal agreement was therefore all the more to be regretted. At the Geneva summit in November 1985 the two leaders had agreed in principle to a 50% reduction in strategic offensive forces and an 'interim' agreement on INF. These objectives had been acceptable to Reagan because he had achieved his own stated precondition for pursuing serious arms-control negotiations: the modernization of American strategic nuclear forces and a firm start to his SDI programme. And Gorbachev, having begun to consolidate his political power, appeared to be seeking a breathing space in which to pursue his domestic reforms. In addition, the USSR seemed genuinely concerned about the strategic and technical implications of the SDI, which, if

Reagan wished to use it, would provide him with considerable bargaining leverage. Western public opinion had rejected the siren calls of the unilateral disarmers and seemed disposed to put its trust in negotiations, provided that the negotiators showed some signs of being able to make progress.

Developments in the negotiations over the last two years had been modestly encouraging. In late 1986 the Soviet Union had made some concessions on strategic offensive forces and appeared ready to accept the West's conditions for an INF agreement, namely a global and equal ceiling and no compensation for British and French forces. Essentially, the two sides had seemed poised for agreements in these two areas, though some critical issues remained to be ironed out, such as sub-limits to be placed on ICBM warheads and missile throw-weight, and the place of short-range missiles in the INF forum.

The promise inherent in these developments was thwarted by the continuing differences between the two powers over space defences and the future of the ABM Treaty. During the course of the year there were hopes that these differences might be finessed by an agreement to extend the notice to be given before withdrawal from the ABM Treaty from six months to ten years, and, within the Treaty, by allowing fairly broadly defined categories of research. But these hopes were dashed at Reykjavik. Not only could the two leaders not agree; their opposing positions were clarified and solidified. The USSR objected to the development of any space-based defence, and the US sought the right to deploy such a system after ten years. Consequently, any agreement on arms control, even an 'interim' INF agreement, was hostage to a resolution of these publicly declared and prestige-laden differences on SDI. As could be expected, each side blamed the other for failure; yet it is truer to say that both can be blamed. What emerged with complete clarity at Reykjavik was that the differences over the role of space-based defences were just too fundamental to be bridged, arising as they do from competing views as to the best way to promote peace and stability in the nuclear age. The strategic, as well as the political, basis for an arms-control agreement simply did not exist.

Rather than discussing, as they should have done, the basis of an agenda for a full summit in 1987 which offered prospects for an agreement, the super-powers pursued a chimera at Reykjavik: dreams and visions of a world without nuclear weapons and also, for President Reagan, a world with perfect defences. A sense of unreality pervaded the negotiations. That a real opportunity for progress towards a new agreement was lost was made all the more serious because of the declared US intention to breach the limits of the SALT II Treaty, and both super-power research programmes on strategic defence threaten to erode the ABM Treaty. Whether Gorbachev's February 1987 proposal to proceed towards a separate INF agreement will break the stalemate is still unclear. The two sides have agreed in principle to the elimination of all INF missiles in Europe and to a global ceiling of 100 warheads. But differences remain over constraints on short-range missiles and over verification procedures.

Equally serious in the aftermath of Reykjavik were the important repercussions which the lack of progress in arms-control negotiations had for Western public opinion and the North Atlantic Alliance. While Western public opinion rejected Gorbachev's post-Reykjavik attempt to put the onus for failure on the Reagan Administration, there was scepticism that either side had been serious enough. Cynicism, even disillusionment about the negotiations arose, and this may prove more difficult for governments to deal with than the demonstrations of the past. In the longer term, it could lead to a loss of public support for the West's nuclear strategies, modernization programmes and arms-control policies.

The consequent loss of support within European governments for American strategic and arms-control policies, as the details of what happened in Reykjavik emerged, is perhaps more disquieting. Whether or not Reagan agreed in principle to the elimination of all nuclear weapons is not as important as the fact that his vision of strategic defences has now become a serious objective, and no negotiated checks can be tolerated. The conditions that the US had stipulated in February 1986 as a basis for agreeing to the elimination of nuclear weapons – namely the correction of the conventional and short-range nuclear force balances, full compliance with existing treaty obligations, and peaceful resolution of regional conflicts – were apparently forgotten at Reykjavik. European governments were dismayed, for they believe that the West's ability to deter aggression depends – however uncomfortably – on the existence of nuclear weapons. The allies' task now is to ensure that American negotiating objectives are refined to reflect the West's dependence on nuclear weapons and to exert what influence they have towards proposals that are sensible and realistic in terms of the negotiations.

Western Europe's Security Concerns

During 1986 the strategic debate within Europe was quiescent, but lurking in the background was unease and continuing nervousness about the implications of the SDI. The general approach agreed by President Reagan and Prime Minister Thatcher at Camp David in December 1984 – that the deployment of strategic defences would be a matter for negotiation – had been Europe's most important political accomplishment since the President had raised the SDI issue in 1983. But Europe became concerned when the US unilaterally broke out of the SALT II limits despite European objections; sought in the arms-control negotiations with the USSR to gain Soviet acquiescence to American deployment of strategic defences and the ending of the ABM Treaty after ten years; and was seriously considering whether to proceed under a broader interpretation of the Treaty in the interim.

In response, the Western European Union, under the sponsorship of M. Chirac, agreed that further work should be undertaken to refine the substance of the European attitude. The essence of the concern has been that, while Europeans have been prepared to accept the high-flown rhetoric that the world should be rid of ballistic missiles (or nuclear weapons), they do not believe it is in their own interest to support the introduction of any new regime which does not take adequate account of the current preponderance of Warsaw Pact conventional forces in Europe. The possible tensions that could arise both among the European members of NATO and between them (either jointly or individually) and the United States, will need careful management, and may require more extensive consultation within the West.

The Stockholm Conference on Confidence and Security Measures and Disarmament in Europe (CDE) ended with agreement on a package of confidence- and security-building measures (CSBM) which requires notification of certain military activities and provides for observers and inspectors. The measures seek to build confidence by reducing secrecy and to contribute indirectly to reducing the possibility of war through either surprise or miscalculation. The major advance made in Stockholm was that, while the agreements reached at the CSCE meeting at Helsinki in 1975 were at best hortatory, the most recent are mandatory. The Conference was also significant for establishing that hard negotiating can eventually lead to success that is in the interests of all, if there is a willingness to make some compromises to reach an agreement.

The Stockholm measures were regarded by many as insufficient because they do not compel any force reductions. The Warsaw Pact's Budapest declaration, calling for substantial troop reductions in Europe, produced a NATO response in December 1986 proposing a new negotiating forum to discuss stabilization and the reduction of conventional arms from the Atlantic to the Urals. Differences between the US and France over the mandate for the negotiations on reductions and how they would be linked to the CSCE talks prevented NATO from defining its substantive objectives in the negotiations, leaving the initiative once again with Gorbachev.

Renewal of the US/Spanish Treaty governing the maintenance of American bases in Spain is also becoming a significant issue for NATO as the Treaty's expiry date (May 1988) approaches. Under the terms of the Treaty, the two parties must agree to any changes to be made, including renunciation, six months before that date, or it will be renewed automatically. There are signs that there is increasing domestic opposition to the renewal of the Treaty. At the same time, NATO has formed a sub-group of officials to discuss with the Spanish authorities the modalities which would be involved in securing agreement on Spain's military co-operation with the Alliance. It would appear that the Spanish government is content to allow these discussions to take place slowly. Although the March 1986 referendum confirmed Prime Minister Gonzáles' recommendation that the country should remain in NATO, there is little popular support for full Spanish participation in the integrated military structure.

While these and other problems continue to harry NATO officials, the results of the general elections in Norway and the Federal Republic of Germany confirmed broad levels of support for those countries' continuing active membership of NATO and, by implication, continuing support for principal NATO policies. A more severe test of the strength of this support elsewhere will come with the British general election (which must be held by the Spring of 1988). The Labour Party's proposals for the removal of US nuclear bases from Britain, the abandonment of the British independent nuclear capability and efforts to persuade the Alliance that it should adopt a non-nuclear strategy represent a clear departure from present British policies and sit somewhat uneasily with Labour's declared support for NATO membership.

Opinion polls indicate that the British public is wary of this approach and of the effect that implementing Labour's policy might have on Alliance cohesion. At the same time, however, while there appears to be majority public support for retaining a British independent nuclear capability, the Thatcher government's decision to acquire the *Trident* II (D-5) missile from the United States remains controversial (other options, too, lack majority public support). Should the Labour Party form the next government and set about putting its proposals into practice, both NATO and Anglo-American relations will be subjected to new strains.

The New Challenge

In both foreign and internal affairs, Mikhail Gorbachev has been proclaiming that a change in policy is under way. It is too early to say that there has been a significant shift towards greater willingness to compromise or to suggest that the USSR sees its future as likely to be better secured by a rather less antagonistic stance in world affairs, but the tone of its pronouncements, if not yet its actions, has certainly altered. It may be – as it has been all too often in the past – that what is being seen is old, stale wine being poured into new bottles. It may also be shown that, even if Gorbachev and his allies in the Kremlin wish to create new conditions in foreign affairs, the objective realities they face create dilemmas which they cannot evade.

Soviet proposals concerning Afghanistan during the past year were a case in point. It has long been clear that the military action here is a net loss for the Soviet Union. It is expensive in financial and manpower terms; it adversely affects the image of the USSR as a mighty military power; above all, the continuation of the fighting after seven long years damages Soviet international interests, particularly in the Middle East and South Asia. To press on for military success would mean greatly increasing all the above burdens. Instead, Gorbachev would like to end the war through a negotiated settlement and withdraw the Soviet forces from Afghanistan, but he will only do so if he can be sure that the government he leaves behind is a friendly one which Communists control.

As positive steps towards that goal, he replaced Babrak Karmal as head of the Afghan government with Dr Najibullah, a younger, more malleable man, who is being encouraged to offer more inducements to the people. Gorbachev withdrew a small number of surplus Soviet troops from the country with considerable fanfare. He instituted a six-month unilateral cease-fire and called on the rebel forces to do the same. He has suggested that the Soviet Union would look with favour on a coalition government in Afghanistan. He has tried to make it appear that a settlement negotiated by the UN between Afghanistan and Pakistan is just around the corner, and that if a settlement is signed, all Soviet troops would be withdrawn within a shorter time frame than has ever been suggested before.

There are of course barbs within the bait so enticingly held out by Mr Gorbachev. It is clear that the coalition government he wants to see is one in which the rebel elements are in the minority and can be swallowed up easily when the time is ripe. A cease-fire could only work to the advantage of the Najibullah government and Soviet troops. Soviet troops would not be withdrawn until it was certain that no arms or other supplies were allowed into Afghanistan to the rebels. Under these circumstances, it is right for those who support the *Mujaheddin* to view the Soviet proposals with considerable scepticism and for the *Mujaheddin* themselves to refuse to join a cease-fire. Moscow's dilemma is that the ends it desires are incompatible with the means at its disposal. Until it is willing to match the means and the ends, the brutal unproductive war in Afghanistan will continue.

Gorbachev's wide-ranging speech at Vladivostok in mid-1986, an attempt to exorcize the deadening legacy of past Soviet policy in the Far East, also illustrated clearly the difficulties of matching desires and reality. Beneath the careful words his agenda was clear. He would like to improve relations with China quickly, harness Japan's economic and technological strength to help in pulling the Soviet economy out of its weakness, encourage the developing anti-Americanism throughout the Asia/Pacific region by portraying the US as an interfering, aggressive, external power, and ensure a role for the Soviet Union in regional developments. To these ends he offered a plethora of vague, broad proposals heavily laden with expressions of peace and friendship, interspersed with marginal specific ones which attempt to address long-standing Asian concerns. They do not add up to a real change in Soviet policy in Asia. They do, however, provide a more congenial framework within which Soviet diplomacy can manoeuvre in the future, a set of options which were not previously available.

As in Afghanistan, the Soviet Union's difficulties stem from its desire to hold on to past gains (the very gains which have so soured relations with nations in the Far East), while offering only goodwill in return for a change in those relations. China, for example, despite gradual improvement in its relations with the USSR, has made it clear that an end to Soviet support for Vietnam's effort to swallow Kampuchea, a withdrawal of Soviet troops from Afghanistan, and a reduction of the number of Soviet troops on the Sino-Soviet border would be necessary before Party relations could be mended and a true rapprochement achieved. Although he edged forward on the second and third points, both in his Vladivostok speech and in other actions throughout the year, Gorbachev did not offer anything on the first one, which for China is the primary obstacle. Chinese suspicions have not been allayed enough to make the breakthrough that Gorbachev desires in Sino-Soviet relations.

Within China, the dismissal of Hu Yaobang as Secretary General of the party on 16 January 1987 dramatized the difficulties faced by Deng Xiaoping in his extraordinary attempt to restructure China's economy and society. Ostensibly it was Hu's failure to curb growing student demonstrations in favour of increased freedom of thought and speech which led to his enforced resignation. In fact it was a result of the continuing tension between the efforts to liberalize economic planning and execution and to maintain tight control on political reforms.

The need for economic change seems to have been accepted by all China's leaders; the basic argument now is about how far and how fast the changes should go, and how to prevent 'bourgeois liberalization' from transforming the central role of the Party in Chinese socialism. Hu was in the forefront of those pressing for reforms; having left himself vulnerable to those who refuse to accept any political liberalization as part of the process of economic reform, he ultimately had to be sacrificed by his mentor, Deng Xiaoping, for the sake of the reform programme as a whole. With the political initiative having passed, if only temporarily, to Deng's conservative opponents, there will be, once again, a pause in the drive to liberalization in China, and the possibility exists that some reversals may even result.

In Japan, too, Gorbachev's initiatives were treated with great caution. The government of Prime Minister Nakasone, who began a fifth year in office during 1986, would be receptive to an improvement in Japanese–Soviet relations, but only if its basic requirement – a return of the Northern Territories – is met. Gorbachev sent Foreign Minister Shevardnadze to Tokyo in January 1986 (the first visit on this level for ten years), for the first time explicitly recognized Japan's status as an economic great power and world leader, and offered to make a personal visit to Japan; even so, Soviet unwillingness to compromise on the territorial question remains a substantial obstacle to much change in the relationship. The summit meeting, planned for January 1987, was postponed indefinitely because Gorbachev would not meet Nakasone's condition that he should offer new proposals on the Northern Territories. Despite Gorbachev's Asian 'charm offensive', the Japanese conviction that the Soviet Union is its only direct security threat remains unshaken. Agreement within the Japanese government that the defence budget could exceed the hitherto sancrosanct level of 1% of GNP stemmed not least from this view of the USSR.

Nor did other Asian nations rush to applaud Gorbachev's calls for a disarmament conference, for non-nuclear zones in the Pacific and Indian Oceans, or for withdrawal of US forces in the Philippines in exchange for withdrawal of Soviet forces from some unspecified base. The present Filipino government accepts the continuance of the US military bases – albeit with some reservations and a desire to adjust the agreement when it runs out. This may mean higher costs to the US and greater recognition of Filipino sovereignty, but it is a condition that is negotiable.

Much more troublesome were the indications of the fragility of the Aquino government. Throughout the year Mrs Aquino was buffeted from both left and right. Supporters of the previous Marcos regime combined with dissident elements in the military to mount a number of armed challenges to her government; these were unsuccessful, because of the support Mrs Aquino received from loyal military commanders, particularly the Armed Forces Chief-of-Staff, Gen. Ramos. The cost of this support, however, was an inability on her part to meet the demands of the left even half way. Squeezed on both sides, Mrs Aquino's survival was dependent on her popularity with the people: a popularity demonstrated anew in the highly positive vote in the referendum on a new constitution in the first days of February 1987. But there are ominous signs that this popularity could wither rather quickly unless it is fed by means of more rapid economic and political improvement in the lives of the people. Mrs Aquino will need to show more results if she is to survive the coming year. If she does not, chaotic conditions can be expected in the country, and these will threaten the effort to strengthen democracy in the Philippines and revive the threat to the US bases there.

Interlocking Security Problems

The linkage between Western strategic policies in the Atlantic and Pacific theatres was strengthened during the year not only by Prime Minister Nakasone's willingness to co-operate with the Reagan Administration, but also by common trade-policy problems which carry security implications. The dangers posed to friendly political relations by protectionism have long been the subject of debate in both arenas, not least in the context of Japan's growing trade surpluses with Europe and the United States.

During 1986 a new dimension of this problem became apparent. Protectionism threatens not only the quality of political relations but also the economic foundations of security. Poor states cannot afford adequate, modern armed forces, nor do they have the resources to ease acute domestic problems and frictions which can shatter national cohesion. The security that has been built up in the newly-industrialized countries of east and south-east Asia, for example, is founded partly on their prosperity. Hence threats to that prosperity also jeopardize security in a way which is all too often not understood, or is ignored, by those in other states who call for protectionist measures. In the cases of developed states, including the European allies, protectionist exclusion from markets reduces their capabilities to bear a larger share of the joint defence burden. Although during the past year these issues have begun to appear more prominently in the political debates of most trading states, from Western Europe to the South Pacific, they have yet to have a noticeable impact on those American and Japanese politicians who support increased protectionism.

Anti-nuclear sentiment has also become a problem throughout the Pacific during the past year. Under the Treaty of Raratonga, establishing the South Pacific Nuclear Free Zone, which came into effect in December 1986, Australia, New Zealand and the 13 other signatories are seeking to constrain the nuclear powers which use the Pacific (particularly France and the United States) to abide by a new code of conduct if they do not want to prejudice friendly relationships. The Treaty calls upon the world's nuclear powers to refrain from locating, testing or using nuclear weapons within the zone (although neither the Treaty nor its protocols prevent the transit of warships in international waters). US reactions to the Treaty have been conditioned by concern at the upsurge of anti-nuclear feeling in the Pacific as well as in Western Europe. In refusing 'at this time' to sign any of the three protocols to the Treaty, the Reagan Administration was also sending a message of firmness to allied political leaders espousing policies which would restrict existing American rights to deploy nuclear weapons abroad. The Soviet Union, which has used the Treaty as an opportunity to appeal to public opinion in the Pacific as an upholder of peace, declared its willingness to sign the Raratonga protocols. It also concluded a fishery agreement with one of the most stridently anti-nuclear states of the region, Vanuatu, thus for the first time gaining shore facilities in the South Pacific.

The deployment of *Tomahawk* nuclear-capable cruise missiles by the US Navy has stimulated debate in the western Pacific. Although the deployment does not represent any change in the US Navy's mission in the Pacific, its coincidence with increased public discussion of US maritime strategy has reinforced criticism in Japan, the Philippines and Micronesia regarding US bases there. The critics remain a minority, although they are becoming increasingly vociferous. Nonetheless, given the delicacy of the situation in the Philippines, and the strength of nationalist opposition to continued American use of Subic Bay and Clark Field, US naval policies, particularly on nuclear issues, will need sensitive handling in the coming year if they are not to be counterproductive.

Other Regional Issues

The Reagan Administration, in partial justification of its policy of trading arms both for influence with the so-called moderates in Iran and for the release of American hostages, has claimed that the amount of arms sold to Iran was not enough to cause any change in the balance of power in the Gulf War. In a narrow sense this is probably true, although there was some evidence that the Iranian offensive towards Basra in the early months of 1987 benefited from the infusion of US missiles and spare parts. The greater impact, however, was undoubtedly psychological, although it is more difficult to define. Iran still has a long way to go before it can achieve its aims in the seven-year war, but its efforts have been boosted by the American actions. Moreover, Iran does not need to sweep across Iraq to Baghdad to claim victory; a change in the Iraqi government would suffice, with Saddam Hussein sacrificed to install a friendlier regime. This cannot be in US interests, since a triumphant Iran might pose a considerable threat to the moderate Arab regimes in the immediate vicinity – Kuwait, Bahrain and Saudi Arabia. Yet the United States is now in a difficult position, with the credibility of its policies of neutrality brought into question. Having now little reason to expect constancy from American policy, the Arab Gulf states may think that they, too, might best be served by some dealings with Iran. Indeed, Saudi Arabia had already moved closer to the Iranian position on oil prices even before the scandal broke.

The revelations have also for the time being removed the basis for any leverage which the US might have been able to exert in the search for an Arab–Israeli peace settlement. Hosni Mubarak, whose domestic position has been weakened by Egypt's economic difficulties and opposition from fundamental Islamic forces both inside and outside the country, can be expected to be more careful in his dealings with Israel and even the US, while continuing to rely on US economic aid. Syria, which might have been rethinking its policies in the light of its failure to prevent utter anarchy in the Lebanon, and of British and US efforts to isolate it for its support of state inspired terrorism, will feel justified in its backing of Iran and its confrontational regional stance. And Israel, already less inclined to compromise under Yitzhak Shamir than under Shimon Peres, was angered by the efforts of the White House staff to shift the blame for the Iran/Contra fiasco onto it. US–Israeli relations are too fundamentally important for a split to be allowed to develop, but the climate that has evolved probably precludes any possibility that in the near future the US could compel the Israeli government to take any action it did not already want to take.

The scandal has also seriously undercut Western efforts to deal with the twin problems of terrorism and hostage-taking. It has long been clear that the only effective measures are a combination of forceful action when the terrorists can be identified (as in a hijacking) and determined refusal to give in to blackmail when hostages are seized and their location cannot be identified. Until all states whose nationals are now subject to these actions make it clear that every terrorist demand will be fruitless, they will be faced with expanding demands. There may always be some religious or political fanatics, or even a few psychopaths, who will carry out terrorist actions, but state-backed and state-financed terrorism – the most dangerous variety – will not disappear so long as there is a useful dividend.

In public, in the earlier part of last year and in years past, the Reagan Administration had been in the forefront of the effort to convince all Western countries who are victims of terrorism that concerted efforts directed at refusing to accede to demands was the only sensible response. It will now be difficult for the US to take a leading role in the necessary re-establishment of a firm and credible policy on this question. Without an agreed position of 'no surrender' it will be the terrorists who are in control, not Western governments.

Given its credibility problems in the wake of the Iran/Contra revelations, it may be that the Administration will now find it best to rely on the Latin American nations of the Contadora group to rejuvenate their proposals for ensuring that Nicaragua's government accedes to international norms prohibiting the export of revolution. To be sure, this will not meet the more recent Reagan policy aim of bringing about a change in that government, but it will fulfil the originally announced goal of constraining Communist ambitions in Central America. Even with the additional funds and arms which the US supplied to the Contras, there was little indication that they would be capable of overthrowing the Sandinista government in Managua; after recent developments it is doubtful whether the Congress, now controlled by the Democratic Party, will be prepared to pay the continued heavy costs of the Contra force. And the Congress can be expected to keep a sharper eye on private efforts, particularly of the kind that the Iran/Contra episode revealed to have been encouraged by the White House staff. President Reagan, in his State of the Union address, has made it clear that he plans to put forward a request for further funding, but if this fails in the Congress he will have to adjust his policies to fit the new realities.

The Congress had already forced the Administration to adapt its policies towards South Africa earlier in the year. The US has now imposed sanctions and dropped its efforts to encourage reform of the apartheid system through 'constructive engagement'. However, neither increased pressure from governments, nor that from Western businesses (which left South Africa in large numbers during 1986), nor yet violence by restive Blacks in the townships have moved Pretoria to temper its harsh policies. If anything they have had the opposite effect, creating a *laager* mentality and resulting in heavier repression in South Africa and new pressures on its neighbours.

Disaffection on the right wing of his ruling National Party and the unhappiness of tradition-minded White South Africans have prevented State President Botha from proceeding with even the limited reform measures which he had hoped would be acceptable to moderate Black leaders and his own followers. The reimposition of the State of Emergency in mid-1986, harsh police and army action, the arrest of large numbers of Black leaders and activists, and a tightening of press laws to prevent the spread of information leading to sympathy action against the government have dampened overt expressions of opposition throughout the country. But problems are multiplying. The economy has been further weakened; in 1986 White emigration was greater than immigration; the National Party has begun to fracture; and Black violence, smouldering beneath the lid imposed by the security forces, could explode with greater violence at any time. With extremists coming more to the fore on both sides of the racial divide, the prospects of any possible compromise have long since grown dim, and it is difficult to foresee anything but further deterioration in an already parlous situation.

Conclusions

One of the key questions to arise out of 1986 concerns the nature and extent of the reforms that Mr Gorbachev has begun in the Soviet Union. There can be no wholly unambiguous answer, since it is simply too early to tell. Indeed Gorbachev himself probably did not expect an immediate positive return from the new policies he outlined during the year. To bring about change, particularly in long-entrenched positions, it is often necessary first to create a new climate of opinion that favours change. Deng Xiaoping is reported to have said that it will take at least four years before one can judge whether Gorbachev is serious about real change in Soviet attitudes and policies. But it does look as though he has begun the process with vigour and commitment. What the West can do is to encourage this process in those areas where its own interests will also be served.

One such area is that of arms control. This remains the focal point of East–West relations, and any improvement in those relations depends on agreement on arms-control questions. Some useful work was done before the Reykjavik summit meeting, and the compromises that were hammered out there remain as a partial foundation upon which to build. Both leaders have good reason to seek an agreement. It would give Gorbachev a solid accomplishment to enhance his own stature as a world leader and that of the Soviet Union as a world power, and it would reduce the possible need for heavier investment in a new arms race which would impede his plans for the Soviet economy and his hopes of raising living standards. For President Reagan, negotiating a sound arms-control treaty could prove to be of great help in rising above the after-effects of the Iran/Contra affair and regaining the support of Americans and Europeans.

The prospects of concluding such a treaty in the next year have improved so far as an INF agreement is concerned. But on strategic nuclear forces neither side shows much inclination to budge from the fundamental positions adopted in Reykjavik – the one insisting on pursuing defence in space, the other adamant in opposing this. Without some movement here the opportunity that was lost in 1986 may be beyond recovery.

Time is rapidly running out for the Reagan Administration to regain its equilibrium. The long process of electioneering in the US will begin to take its toll late in 1987, as Republicans and Democrats manoeuvre for advantage in the Presidential election due in 1988. Under the best of circumstances, bold action of any kind in foreign affairs would be difficult in the next two years; with the President weakened and preoccupied by the fall-out from the Iran/Contra affair, it is even more unlikely. Nonetheless, there are significant problems to be faced. The confidence of America's allies in US leadership must be recovered, and the threat of a trade war averted. New initiatives by the Soviet leader must be dealt with. Fledgling democracies in the Third World must be helped to take wing. Unless President Reagan can regain his grip on authority and exercise the full power of his office, 1987 and 1988 seem destined to be years of drift.

Strategic Policy Issues

THE CHALLENGE OF DEMOCRACY

As a result of one of those combinations of events which sometimes lend shape and even the appearance of purpose to the affairs of men, the 1980s have seen an extraordinary revival of interest in the current fortunes of democracy as a method of governance, and in what role, if any, it should play in the conduct of foreign affairs. Since 1975 Spain, Portugal and Greece had been restored to the community of Europe from the limbo of dictatorship, and India had reclaimed the honour of being the world's largest democracy. By 1985 democracy's advance had speeded up, most notably in Latin America; of two despotic dynasties there, Somoza's in Nicaragua had already come to the end of its life, and Duvalier's in Haiti was running out of steam.

Military juntas, too, were surrendering political roles they were no longer able or willing to justify: Argentina regained democracy in 1983 and Uruguay in 1985. In Bolivia, the new democratic government, which had succeeded eighteen years of military rule, not only held elections in July 1985 but turned power over to the opposition. In the same month, the government in Peru became the first democratically elected government there since 1912 to complete its term without military intervention. In November 1985 there was the first election of a civilian president in Guatemala for fifteen years. In 1978 only four countries of Latin America had had democratically elected governments; by 1987 only two – Chile and Paraguay – did not. US Secretary of State Shultz could claim in early 1986 that the growth of the 'democratic center' and the decline of 'political oligarchy' in Latin America had brought 90% of the people of the area under democratic government, compared with only one-third a decade earlier.

Early in 1986 the sudden demise of two notorious authoritarian regimes captured world attention. On the same day that Duvalier fled Haiti – 7 February – President Marcos held an election in the Philippines which he hoped would help him fend off pressure for reform. In the event, it precipitated the first internationally televised revolution in history. The US Administration, which could hardly claim to have willed, much less inspired, these turns of events, made the best of the situation. Having facilitated the peaceful departure of Duvalier, it helped President Marcos, too, to face the always disagreeable task of coming to terms with the inevitable. By accommodating itself to developments, Washington found itself rewarded by having backed the winning – and popular – side in both cases.

Given the frustrations of diplomacy, tactics which work are sometimes raised to the dignity of strategy. In his address to Congress on 14 March, President Reagan associated the victory of democracy with the aims of US foreign policy: it would be America's intention, he declared, to 'oppose tyranny in whatever form, whether of the right or the left', a commitment quickly dubbed the Philippine Corollary of the Reagan Doctrine. On the same day the UN Commission on Human Rights adopted a US resolution condemning human-rights violations in Chile – a country which President Reagan bracketed with Cuba, Nicaragua and Paraguay as one of the remaining non-democratic regimes in Latin America. Moreover, the new US ambassador to Paraguay made it known that his brief was to help to promote contact between the Stroessner regime and its opposition.

As President Aquino abrogated the constitution and assumed legislative powers to bring her revolution under control, President Chun Doo Hwan of South Korea stressed the difference between the two countries – but he observed also that events in the Philippines could hold 'a lesson for us', and began to negotiate with the opposition on constitutional reform. No doubt other authoritarian leaders, too, pondered the implications of what had befallen one of the most durable and seemingly secure of their autocratic fellowship.

Putting American foreign policy at the service of democratization met with a mixed reaction. Some, perhaps cynically, suggested that such a commitment might better be considered in the context of the President's efforts to win support for his Nicaraguan policy, rather than *vice versa*. Against those who claimed that the President's 14 March speech was a radical departure, there were others – Admiral Poindexter among them – who professed to find nothing much new in it. Both, perhaps, had some grounds for their views, for, as one observer pointed out, the Reagan Administration had shown a '150-degree, if not 180-degree, change' in its approach to human rights since its early days, when it had alleged that the Carter presidency's concern for such matters had undermined US strategic interests. By endorsing Ambassador Kirkpatrick's thesis that a distinction could and should be drawn between totalitarian Communist regimes and those of a more traditional authoritarian colour, the Reagan Administration had indicated that policy would not be deduced directly from US historical political principles, and support for US confrontation with the Communist world would be welcome even from states which did not share them. But as the years passed, and under the pressure of both events and public opinion, human-rights considerations came to play an increasingly important part in the thinking of the Administration. It remained only for the Philippine crisis to demonstrate that, at their best, autocratic allies are useful only so long as they can deliver stability; at their worst, they are apt to intensify the forces of disorder preparing for their eventual departure from the scene.

The problem with doctrines, or their corollaries, is that they tend to address a kind of problem which in the real world seldom gets posed in quite the way it was expected to be. In the months that followed the President's 14 March speech, qualifications had to be made to the scope of the new policy: Administration spokesmen pointed out that it was not their aim to replace bad governments with worse ones, and that a democratic alternative was not always available in every case where a non-democratic regime could be identified. Such deference to the art of the possible was also detectable in the vacillation of Administration policy towards Chile – one instance, at least, where the United States might have been expected to stoke up the pressure for democratic change. The Pinochet regime was indeed criticized for its depressing record on human rights, but it was also supported in its attempts to raise credits from the IMF and international private banks. The Administration also yielded to increasing pressure by adopting sanctions against South Africa, but it could hardly have been said to have taken the lead in the search for a democratic solution to this country's problems. During 1985 the US Congress had made a *volte-face* by agreeing to 'humanitarian aid' to the Nicaraguan Contras, voted funds for the resistance movements in Kampuchea and Afghanistan, and lifted a ten-year ban on aid to Angolan rebels. Although defended as support for those resisting undemocratic regimes, such selective initiatives looked much more like a continuation of established Administration policy than a portent of a new and disinterested campaign to advance the democratic ideal as such.

The year-end balance sheet, therefore, was still inconclusive; it remained to be seen whether the centrality of East–West relations would continue to monopolize and define all US strategic interests or whether some foreign-policy issues could be set aside for consideration on their own merits, as judged by historical American political principles; whether Project Democracy was a new version of President Wilson's Crusade or an old device for covering political operations in pursuit of White House objectives. Such uncertainty may well account for the failure of the President's initiatives to stimulate interest among America's principal democratic allies, none of whom have indicated a comparable belief in the merits of this method of promoting democracy in the world at large.

The Democratic Camp

Any attempt to assess democracy as a touchstone for foreign policy encounters difficulty from the start, for government of, by and for the people is neither in theory nor in practice so simple as many of its partisans would like to believe. Democracy has its own lore, its unique culture, which societies draw upon in conducting public affairs. Despite all the passions which politics can inspire in a democracy, for example, both winners and losers in the struggle for power know that victory is never absolute, nor defeat final; history never comes to an end, and politics go on forever. Despite the freedom which democracies provide for religious belief – or perhaps because of it – they nonetheless seem to have found accepted strategies for determining what must be rendered unto Caesar and what need not. Above all, the democratic spirit holds that, while power may grow out of the barrel of a gun, legitimate power can be found only in the garden of legal process, however rank with weeds that plot may sometimes appear to be.

If definitions seem elusive, we might assume that we can at least recognize democracy when we see it. But this, too, is not always so easy. Stalin took the view that any government which was not fascist was democratic – but then he was not over-familiar with the latter, either in theory or in practice. The very word 'democracy' has in fact become a generic designation, not protected by trademark: we are all democrats now, as Edward VII might more accurately have said. Nomenclature often misleads: some of the monarchies are among the most democratic of governments, but calling a state a 'People's Democracy' is usually a clear sign that it is no such thing. Nor are voting statistics much of a guide: in non-democratic states, the electoral process seems to serve the same purpose as feudal homage – a mechanism designed to confirm and consolidate authority, not pose a threat to it.

All the same, it seems reasonable to say that some states in the world are clearly democratic, and some are clearly not. In between lies a range of governments less easy to characterize. Some societies are not viable political entities to begin with; some are not secular states; some are too poor to support adequate civic life; and some are victims of systematic corruption masquerading as government. Some contain a mixture of elements which renders classification a matter of argument. Should the name of democracy be bestowed, for example, on South Africa, which provides for an opposition party, an independent judiciary and the mechanics of free elections and yet excludes the majority of its people from the political process? Was it an attachment to this form of government, or to something else, which prompted State President Botha's remark that democracy is 'much too valuable to be shared with everyone'?

A comparison between democratic states and others points to the European origin of the democratic idea and indicates that democracy is still very much a minority system, for it does not provide the means of government for most of the world's population. The fifteen most populous democratic states contain an estimated 930 million people, compared to some 1.9 billion living in the fifteen most populous non-democratic states – and the discrepancy would be still greater if the lists were extended to include the smaller countries. Nonetheless, if fewer people live in democracies there is no doubt that they command a large preponderance of the world's wealth. Of the top twenty states ranked by GDP, all but seven are democracies, and the ten wealthiest democracies have a combined GDP roughly four-and-a-half times that of the wealthiest ten states in the non-democratic camp. It is hard to avoid the conclusion that, whatever other advantages it offers, democracy seems to be the form of society most likely to accompany long-term economic development and material progress.

From Ataturk to Zia, autocratic leaders of developing countries have felt that democracy is not suited to the special needs and historic handicaps which face their societies. The unconscious irony in the expression 'guided democracy' reflects the frequently-observed paradox of modern political forms imposed upon societies which live by ancient traditions of very different lineage. Faced with the taxing variety of human social life, even the most uncompromising partisan of the democratic model will hesitate to generalize about the universality of its application. Many countries today, particularly in Asia, are passing through a period of rapid industrialization, burgeoning prosperity and social transformation without – so far as one can see – any growing demand among their populations to control their own political destinies. Elsewhere, in Africa for instance, popular political consciousness is still directed to the issues of colonialism, or else attenuated by tribal or other parochial loyalties not always consistent with the habits and skills of civil democracy. As developing nations develop, government from the top down will often prove less and less able to cope; the competence of its economic management is challenged by a more powerful and sophisticated market-place, and the legitimacy of its rule called into question by a rising political consciousness. In one-party states, divisions within the ruling party can emerge which approximate – in somewhat distorted form, but with no less partisanship – to the rivalry of competing political centres. This in turn can precipitate a level of pre-democratic political activity – from which the population and the 'stability' of the state is often saved by the intervention of the military, thereby starting the process all over again from square one.

For such a recent movement in history, democracy's success has been dramatic. It effectively replaced monarchy in practice, if not in theory, throughout Europe and was exported to the rest of the world as a by-product of colonialism. Although many even of its partisans once seemed to feel that it was too good to last (think of such works as *Brave New World*, *When the Sleeper Wakes* and *1984*), democracy not only triumphed in its war with fascism but has flourished handsomely during its long containment of Communism. It has become a traditional assumption that the spread of democracy throughout the world will promote security and establish peace among nations. Thus President Reagan in 1986: 'The surest guarantee we have of peace is freedom and democratic government', echoing President Wilson's blunter thesis of almost seventy years before: 'A steadfast concert for peace can never be maintained except by a partnership of democratic nations . . . only free peoples can hold their purpose and their honor steady to a common end and prefer the interests of mankind to any narrow interest of their own'. If the democracies of the world can be made safe from their enemies, they will be safe from each other.

On the face of it, however, there seems no reason to suppose that any of the usual sources of dispute among nations – boundaries, irridentist claims, historic rivalries, religious enmity and commercial competition – would not be capable, in certain circumstances, of turning one democracy against another. We would not expect, for example, that a mutual commitment to democracy would, by itself, resolve the quarrels between Greece and Turkey over the waters which lie between them, nor yet that the form of government on either side of the Pacific basin had much to do with the conflict of strategic interests there between the United States and Japan in the 1940s. Indeed, it was a concern to prevent yet another falling out among the post-war democracies of Western Europe which led many European leaders, together with the United States, to move towards the construction of a United Europe.

Even so, the testimony of our lifetime (to the extent that it yields any conclusions) does seem to suggest that major threats to world peace arise from the ambitions – or fears – of non-democratic governments in pursuit of ends which transcend the ordinary civil aspirations of the populations they control. The great democracies have always proved to be reluctant warriors. It was not popular enthusiasm, for example, which propelled the United States into the wars it has fought in this century; but popular opinion did play a part in limiting its involvement in one (Korea) and ending hostilities in another (Vietnam). A democratic government is by no means immune from rivalry with its neighbours, but at the very least it will have to evaluate the importance of such rivalries against the priorities which most electorates assign to securing a better life for themselves and an even better one for their children. The existence of nuclear weapons clearly sharpens such popular pacifism.

History alone may not even be a definitive guide on this question, for we must also allow for the changes in social life and the growth of political sophistication which have been among the significant achievements of political democracy over the past fifty years. The growth of multinational firms has created a shared economic infrastructure among the advanced democracies; a revolution in travel, together with an explosion of information through electronic media, have contributed to a tendency towards increasing co-operation and mutual understanding.

The Necessity Of Choice

As 1986 drew to a close, there was ample evidence – if evidence were needed – that new democracy is often fragile and must fight to maintain itself. The Marcos regime may have been both defeated and discredited, but the legacy of its mismanagement and corruption was an impoverished, volatile and divided society which the Aquino government will need much skill and luck to put right, despite the massive endorsement it received in the February 1987 referendum. In Ecuador, democratic government was at risk from the menace of terrorism, in Argentina, from the memory of it. At the end of a year which had shown so much promise as it began, there was little progress towards democracy to report in such countries as Paraguay and Chile, while throughout other parts of the world civil war for the control of one-party states seemed to be the pattern, rather than a ground swell in favour of democratic reform.

In consequence, President Reagan's 14 March speech can be regarded as a rhetorical gesture, an afterthought, designed not so much to chart a new course for US foreign policy as to raise the Reagan Doctrine from just another cold-war dictum to something of more general and historic significance. Although the Philippine Corollary was described at the time as 'potentially the most interventionist foreign policy since Woodrow Wilson proclaimed a crusade to make the world safe for Democracy', it is likely to find no more application in the real world than did the Monroe Doctrine in its day or, for that matter, President Wilson's Crusade in his. Democracies can be overthrown, and – even where democratic traditions are strong – from time to time they are. Unlike fascist, Communist or non-ideological dictatorship, though, democracy cannot be imposed from above or outside.

But if democracy cannot be imposed, it is often possible to undermine – or, conversely, prop up – an autocracy. In the last analysis, it is this option which the major democracies must relate to their security interests, as they perceive them, and it is the manner in which they do so which has sparked off the most debate. Non-democratic regimes obviously come in many forms and often (like the present leaderships of Iran or Libya) command considerable popular support, which their reliance on repressive methods of government would seem to belie. Some possess the means to threaten the security interests of the major powers, whether because of their ability to undermine peace in their region, because of their control of natural resources, or simply because of their geographic position. Within the democracies, controversy has always surrounded their relationships with friendly non-democratic states; but, ironically, just as much has arisen over action taken by democracies against unfriendly ones. Even the friendly non-democratic regimes present their democratic allies with a niggling worry that, sooner or later, they may well be replaced by unfriendly democratic ones, only too ready to renounce the alliance that helped to keep their predecessors in power.

It is normally the hope of the great democracies to help autocratic allies prepare for a peaceful transition to democracy where that change appears inevitable. Such a scenario, however, usually involves factors which cannot be supplied at will or even fully controlled by anyone – the existence of responsible alternative leadership, the habits of orderly political activity, and, above all, a capacity for compromise and patience on all sides. Situations differ markedly: some countries have only recently lost well-established democratic institutions (Chile); others (such as Paraguay) have little memory of them. Some possess a political opposition, others do not. And, in those that do, the opposition may range from a prospective autocracy even worse than the one it hopes to replace to an array of conflicting and untried forces that would tear the society apart if let loose to campaign.

Judging the pace of reasonable change will usually be a challenge to all concerned – and they are liable to get little thanks for whatever determination they are able to make. In the end – as in the beginning – the most important work in facilitating the transition to democracy will be to convince and then to help existing authority to build popular confidence and identification with the state, making it a source of justice and civil order which will survive the eventual change in the form of government under which it is managed.

The desire for democratic systems continues to be a force in the world. Towards the end of 1986 it returned to figure importantly in discussion of international affairs – not this time in the Third World, but, surprisingly, in the very heartland of non-democratic ideology. Nationalist rioting in Kazakhstan paralleled strident and explicit demands for democracy in China by student demonstrators. As 1987 began, Mr Gorbachev startled the world, and no doubt many of his colleagues too, with a call to bring democracy into the Soviet Communist Party in the form of secret ballots and multiple candidates for certain low-level posts. Such unexpected proposals – whatever their ultimate intention or likely fate – could not but prompt a new line of speculation about the significance which such developments may have for the future.

The Philippine Corollary had raised fundamental questions in the minds of some about what foreign policy should be, and what aims it should serve. George Kennan had argued persuasively before the President's 14 March speech, as did Henry Kissinger after it, that a commitment to promote democracy such as the President had made was a departure from the traditional approach to diplomacy, which seeks to derive policy from a calculation of national interest, rather than from an expression of moral sentiment. But the paradoxes of policy often make the political leadership in a democracy, and public opinion, feel uncomfortable. They feel that change is in the nature of things, they see events moving in some direction and assuming some new coherence which they expect government to anticipate and relate to the values and needs of their society. Events of the past year may therefore suggest to some that the world has entered a period of transition – that a shift in political geology, long pent up, is finally under way.

For almost forty years, the West had resisted what it perceived as Soviet expansion, replying with measures ranging from containment to the unrecorded and inconclusive skirmishes of covert action. The weapons of externally controlled political subversion and insurgency, once much valued by Soviet ideologists and feared by their counterparts in the West, were shown in the long run to have been greatly overestimated by both: in the long run, in fact, the enemy within is proving to be a far greater a problem for the Communist world than for its opponents.

As Communist political influence declines in the West, and Soviet-backed insurgency is increasingly overcome in parts of the Third World, the Communist camp itself is facing increasing pressures from unreconciled national and religious loyalty, and the accumulated resentments which a monopoly of power always engenders among those who do not share it. Serious design faults have shown up in the Communist machinery of government: it has proved dangerously unstable when required to perform the basic governmental function of transferring power, but it is often glacially immobile once the players and the policies of a new settlement are put in place. In the Soviet Union's immediate neighbour and client state, Poland, massive disaffection forced the abandonment of civilian government and the imposition of martial law, an extraordinary and unparalleled breakdown of authority among European nations. Among the outposts of the Soviet empire – South Yemen, Angola, Kampuchea, Mozambique, Ethiopia – Communist allies encountered civil disorder or insurgency they could not always control and, in Afghanistan, a resistance which Moscow seemed unable to defeat even with direct military intervention.

For the moment, and no doubt for quite some time to come, the significance of recent events will be hard to judge. The Chinese students have returned to their studies, their outburst seeming to have only stiffened resistance to 'Western' notions of reform. Gorbachev's initiative remains to be judged by its consequences. These, we can be fairly sure, will not include a Kremlin-led movement for political democracy as the West knows it, nor even the extension of political power beyond the Communist party. On the contrary, Gorbachev seems to regard democratic reforms in the Party structure as a means of enhancing its effectiveness as a political oligarchy: not so much socialism with a human face as socialism with a Western capitalist efficiency.

Gorbachev appears confident that he can exploit the practical advantages of democracy with impunity, but those who believe in democracy may have their doubts that it will prove that easy. As Professor Kolakowski points out, it is not possible for a society to be neutral about its own values. Many in the West are therefore already beginning to wonder if we are about to see salami tactics operating on behalf of democracy for a change. In Seymour Martin Lipset's definition, democracy 'is not only, or even primarily, a means through which different groups can attain their ends or seek the good society; it is the good society itself in operation'. General Secretary

Gorbachev may eventually discover that it is not only a technique of enlightened management but a theory of legitimacy of the state. To the extent that this theory finds new adherents in the world, the effects must cause us to recognize further signs of fundamental change in international order.

THE FUTURE OF US ICBM

Throughout this decade, modernization of the ICBM force has been one of the most controversial issues in American strategic policy. In 1986 a milestone was reached with the deployment of the first MX (or *Peacekeeper*) missile in a *Minuteman* III silo. Nonetheless, a number of critical decisions remain, and the future of US intercontinental ballistic missiles continues to be at the forefront of the US strategic policy debate.

The history of US ICBM modernization in recent years has been a chequered one. In 1980 President Carter recommended the deployment of 200 MX (each with 10 warheads) in hardened multiple protective shelters (MPS). The plan called for the missiles to be ferried on transporter trucks between 4,600 concrete shelters on a road network of some 8,000 miles. The Congress had problems with this concept, but before it could act President Reagan cancelled the MPS scheme in January 1981, partly in response to political opposition from the proposed site areas. But he remained committed to the MX itself, and to finding a basing mode that would reduce the vulnerability of American silo-based ICBM to attack by the increasingly accurate Soviet ICBM.

As a temporary expedient, the Administration first proposed basing the MX in existing *Minuteman* III silos and later recommended a basing scheme involving closely-spaced silos. In late December 1982 Congress, dissatisfied with the Administration's proposals, blocked further MX procurement until a suitable basing mode could be found. In an attempt to break the log-jam, the President appointed a Commission on US Strategic Forces (the Scowcroft Commission) which recommended that the US ICBM requirement should be met by a combination of small single-warhead missiles (which would provide greater survivability for the force), and a reduced MX deployment of 100 missiles in fixed silos.

Congress ultimately approved the general outline of the compromise but authorized only 50 MX in fixed silos, withholding approval for the second 50 until the Administration could devise a more survivable basing mode for them. At the end of 1986 the Administration announced its intention to proceed with the development of 500 single-warhead mobile missiles (the *Midgetman*) and proposed that the second 50 MX be based on rail cars.

The Role of ICBM

The importance of the continuing debate over US ICBM stems from the fact that, in the short term, US nuclear policy must continue to be based on a strategy of nuclear retaliation. (Whatever the long-term prospects for strategic defences, they cannot be expected to provide a genuine alternative basis for security until at least the twenty-first century.) To be successful, US retaliatory forces must be able to survive an attack and threaten unacceptable damage, whether in response to a massive Soviet attack or a more limited one. But the question of how to assure a credible defence has become more complex in recent years, as the strategic arsenals of both super-powers have expanded beyond the traditional elements of the strategic triad (bombers, ICBM and SLBM) to include cruise missiles, which are being deployed on land, aircraft, ships and submarines. The growing diversity of strategic systems and their increased capability as a result of technological improvement, has created new strategic options for the US; at the same time, however, the threat to those forces has also increased as a result of similar developments in Soviet nuclear forces.

The Scowcroft Commission argued that a triad of survivable ICBM, SLBM and bombers was needed to complicate Soviet attack planning and to hedge against the possibility that one or two of the three legs might become vulnerable. It judged that ICBM continued to be necessary because of their accuracy, secure and rapid command and control, quick flight time, ability to retarget, and their effectiveness in deterring Soviet threats of massive conventional or limited nuclear attacks. The Commission also concluded that a more stable structure of ICBM deployment would exist if both sides moved towards survivable methods of basing and single-warhead missiles.

The principles underlying the Scowcroft recommendations have come under increasing challenge over the past few years. The President's proposals for strategic defences that would render ballistic missiles 'impotent and obsolete' calls into question the desirability of embarking on a multi-billion dollar ICBM development programme. Similarly, the US is seeking to use arms control to achieve major reductions in strategic offensive forces, and ultimately the elimination of all ballistic missile warheads – thus undercutting the arguments for ICBM. It is also calling for a ban on mobile missiles, unless the USSR can define adequate verification measures, and suggesting that in a START agreement it might forgo the possibility of deploying *Midgetman* or MX except in fixed silos. The development of a small, single-warhead, mobile missile has also been criticized; some have argued that, with improvements in mobility, missiles with more than one warhead might prove more cost effective and just as stabilizing as single-warhead weapons.

The issues surrounding ICBM modernization have acquired particular prominence as the Administration presses forward with its plans to develop the *Midgetman* and secure Congressional approval of a new basing mode for the second 50 MX. Policy makers now face several critical questions. Does the US need a survivable ICBM force to assure a credible nuclear deterrent? If it does, should it choose a small, mobile *Midgetman* or more MX in a new basing mode – or both, as recommended by Secretary Weinberger? Will missile defences turn out to be cost-effective in promoting survivability, and what would be their implications for the ABM Treaty? The answers will need to take account of developments in the capabilities of both sides as well as the opportunities which new technologies offer to both offensive and defensive forces.

US ICBM Programmes

The United States is now deploying 50 MX missiles in refurbished *Minuteman* III silos, without additional silo hardening. The MX is a three-stage, solid-propellant ICBM, medium-sized (by SALT standards), weighing about 190,000 lb. It can carry up to 10 Mk 21 re-entry vehicles (RV), and could be fitted with various penetration aids if the Soviet Union were to deploy missile defences. Each RV carries a nuclear weapon in the sub-megaton range and can be independently targeted. The missile uses an all-inertial guidance system (the Advanced Inertial Reference Sphere: AIRS), while the RV is of a high weight-to-drag design – to keep it from being deflected as it enters the atmosphere. Its standard of accuracy is close to the best possible without terminal guidance, giving it an excellent hard-target-kill capability. With an operational range of 5,500 nm, MX can reach virtually all important military and civilian targets in the Soviet Union.

The US is also developing a small ICBM (SICBM or *Midgetman*) in line with the concept recommended in the Scowcroft Commission report. Being designed as a single-warhead missile weighing 37,000 lbs and capable of operational ranges of 6,000 nm, it is expected to begin deployment in 1992. Its design, warhead and guidance system will be similar to those of MX. The Air Force has rejected recommendations that additional warheads should be added to the missile, on the grounds that a three-warhead version would not be sufficiently mobile – in any case, even a two-warhead version would have required a new round of competition for some of the contracts and a two-year delay in the programme.

Mobile and Fixed Basing

Mobility enhances survivability and improves crisis stability, because it increases the number of warheads the attacker must use to ensure that he destroys one US missile. If the USSR must use more warheads than it destroys in an attack (the 'exchange ratio'), its incentive to initiate an attack will in theory be greatly reduced. Using highly accurate ICBM, it may require only one or two warheads to destroy a fixed silo, whereas to destroy an equivalent number of mobile missile warheads it would be forced to barrage a large area.

The number of warheads required depends on how widely the mobile missile can be dispersed – a function of both the mobility of the launcher and the amount of warning time given and acted upon.

But mobility is not the only means of improving survivability. Hardening of silos, deception and concealment also increase the number of warheads an adversary must use to ensure the destruction of a launcher. And, of course, these measures can be combined.

Choosing a Basing Mode

Rail Mobility

The Administration is now seeking approval for the second 50 MX in a 'rail mobile garrison' basing scheme. The missiles would be deployed aboard trains (two per train) located at up to seven garrisons spread across the continental United States. Each garrison would contain three to four missile trains designed to look like civilian rail traffic, so as to minimize the likelihood of detection by photo reconnaissance. Assuming that the rail cars could withstand blast pressure of about 10 pounds per square inch (psi) above normal atmospheric pressure, and that approximately 60,000 miles of high-quality track is available for deploying them, it would take something like 10,000 one-megaton warheads to destroy this system – an exchange ratio of approximately 20:1.

To avoid provoking public opposition to the idea of nuclear weapons travelling around the country's railroads, the trains would be garrisoned during peacetime and dispersed onto the civilian rail network only upon warning. The basing mode would, however, require that steps be taken upon strategic warning (at least hours, if not days) for the trains to disperse effectively. If only tactical warning (15–30 minutes) were available, the trains would be unable to get far enough away from the garrisons to avoid being destroyed by a relatively small attack. (For example, should only tactical warning be available, if the trains left each of seven bases along four rail lines and travelled at 50 mph, it would take only 25 one-megaton warheads to destroy the missiles at a particular garrison – an exchange ratio of approximately 1:3 in the attacker's favour). In addition, the rail-mobile scheme is potentially vulnerable to 'non-standard' threats. Soviet agents might be able to place explosives along sections of the rail networks, to be activated during times of severe crisis, thereby immobilizing this part of the US ICBM force.

Super-hardening Sites

Recent advances in silo construction have made it possible to build super-hard silos that could withstand tens of thousands of psi peak overpressure. If this degree of hardening can be achieved, such silos could survive a nuclear attack, unless it were so accurate that the silo was within the crater carved out by the explosion of the incoming warhead (in which case nuclear-weapon effects – ground motion, rather than blast – would become important).

The advantage of super-hard silos is that they can be relatively close together without the worry that one large warhead could destroy many silos simultaneously. This increases their survivability, because closely spaced silos complicate the attacker's 'fratricide' problem (some incoming warheads perhaps being destroyed by the nuclear explosions of others). Even if the attacker minimizes fratricide by means of a well-timed attack, he still needs to use multiple waves of warheads if he is to achieve a high probability of destroying the super-hard silos. However, the dust clouds from the initial explosions would effectively preclude a second-wave attack until the air had cleared several hours later, because re-entry vehicles could have their trajectories disturbed, or could possibly burn up in the atmosphere if there was too much dirt and dust in it.

Since missiles can be launched through dust clouds more easily than warheads can re-enter the atmosphere through them (due to the missiles' slower velocity at launch), the defender may launch his surviving ICBM in the interval between the first and second attacks. To guard against this, the attacker could adopt another tactic, called a 'pin-down' attack. This involves continually detonating nuclear weapons above the opponent's ICBM field in an effort to prevent him from launching his missiles. But such attacks consume a large number of nuclear weapons, and their reliability is questionable.

When all these factors are taken into account, the exchange ratio for attacking super-hard silos does not favour the attacker. Such silos were rejected for the MX basing scheme, however, because of present and future uncertainties. Estimates of silo hardness depend upon a recalculation of the nuclear effects data which exists from only a few weapons tested in the atmosphere in the 1950s, and the ban on testing in the atmosphere makes it impossible to verify the new results. In addition, even if the silos could be hardened up to tens of thousands psi, they could be rendered vulnerable by future improvements in missile accuracy (i.e. if the warhead were accurate enough to create an explosion crater encircling the silo). Also, although it might take some years, the Soviet Union could deploy earth-penetrating warheads, which would be more effective against hardened silos, and design an attack timed to detonate them simultaneously, thus circumventing fratricide problems.

What to do with Midgetman

Midgetman had provoked considerable debate within the US, but both the Administration and Congress are now supporting full-scale engineering development. The basing mode planned for the missile is the Hard Mobile Launcher (HML): a hardened off-road-mobile vehicle. The current deployment concept calls for 500 *Midgetmen* on military bases either in the south-west United States or in

Minuteman silo fields, or both. Upon tactical warning, the missiles would move off base, and within thirty minutes the HML should be able to disperse over approximately 28,000 square miles. The attacker would then be forced to barrage the entire deployment area to destroy the missile force. Assuming a nominal HML hardness of 30 psi, approximately 9,500 500-kiloton weapons would be needed to saturate this area with sufficient peak overpressure. The Soviet Union now has this number of ballistic missile warheads, and would thus have to use most, if not all, of it to destroy 500 *Midgetmen*. But without arms-control constraints, the situation could change; the CIA has estimated that the USSR could deploy 16,000–21,000 warheads by the mid-1990s. Moreover, *Midgetman* survivability depends upon the critical assumption that the Soviet attack would have to barrage a vast deployment area. This would not be necessary if the USSR could develop surveillance systems to locate each *Midgetman* missile and/ or could develop new manoeuvring warheads to attack them individually. Both responses, however, are technically difficult and far beyond the Soviet Union's current capabilities.

MX, Midgetman or Both?

ICBM basing modes must in the first instance be politically acceptable, and then need to be judged in terms of survivability and cost. Because they involve peacetime basing in garrisons or on military bases, the chosen MX and *Midgetman* basing schemes put a premium on assuring public support. Only in a crisis or upon warning would the missiles be dispersed with their nuclear warheads (in peacetime exercises they could be dispersed without their nuclear warheads, as is the practice for ground-launched cruise missiles and *Pershing* II in Europe). By opting for mobility for both *Midgetman* and MX, the US has sought to improve their survivability. But a determined attacker with a sufficient number of weapons can always destroy land-based ICBM. The question then is whether the basing solutions chosen are 'survivable enough'.

In the past, the US sought to ensure that its ICBM force would survive without any steps having to be taken upon strategic or tactical warnings. Judged by this criterion, the new basing schemes offer no meaningful improvement over fixed-base silos – they will improve survivability of the missiles only if they are able to take advantage of warning to disperse. If it is to justify deployment of mobile systems on survivability grounds, therefore, the US must be prepared to assume it will have at least some warning – tactical warning in the case of *Midgetman*, strategic warning for the rail-mobile, garrison-based MX. In judging whether it is desirable to change the criterion in this way, it is important to keep in mind that other forces – especially strategic bombers – depend on some warning for survival; a misjudgment about the availability of warning (or failure to act upon warning) could seriously erode the survivability of two of the three legs of the triad.

In choosing between ICBM options, cost-effectiveness is essential. If the goal is survivability, the appropriate way of analysing costs is to make assumptions about the number of surviving missile warheads and calculate the cost per surviving warhead. Deploying 500 *Midgetman* on HML was estimated in 1986 to cost some $50 billion over fifteen years (although keeping only a fraction of the missile force on alert would reduce this figure somewhat). During the debate in 1986 some people, including members of Congress and Defense Department Under Secretary Donald Hicks, proposed a MIRV-equipped version of *Midgetman* so as to reduce costs further. They estimated that 500 warheads on a three-warhead version would cost half as much as a similar number on single-warhead launchers, with comparable survivability. Their argument was based on the fact that in a barrage attack (the only means now available to attack mobile missiles), the number of warheads that would survive an attack of a given size depends not on the number of launchers but on the size of the area attacked. However, a MIRV-equipped missile would be heavier, and so probably less mobile; this would limit its deployment radius and therefore its survivability. The Reagan Administration has settled this debate, at least for the initial deployments, in favour of the single-warhead version.

A similar analysis must be applied to the rail-mobile proposal for MX. This missile has the advantages that the US has already incurred its development costs, and that the use of rail (as opposed to a transporter system, like *Midgetman*) reduces the mobility penalty associated with a heavier MIRV-equipped missile. The costs of rail-mobile garrison basing are relatively low – around $2.5 billion to deploy 50 MX, plus another $4–5 billion for the railroad cars and associated equipment The total life-cycle cost to deploy these 50 MX missiles in the rail garrison mode is estimated at some $15 billion. But against these advantages must be weighed the system's dependence upon strategic warning to achieve its projected survivability.

Finally, arms control may be critical to achieving a 'sufficiently survivable' ICBM basing scheme in the future by limiting the offensive threat. If adopted, the Reagan Administration's proposals for a radical reduction in strategic offensive forces and limits on ICBM and throw-weight would enhance the survivability of mobile ICBM. But differences between the two sides with respect to strategic defences have so far blocked agreement. The problem for arms control is that, with the introduction of mobility, decoys and deceptive basing, it becomes more difficult to verify limits on the two sides' nuclear weapons. Mobile basing schemes have caused the greatest concern, since it is difficult to verify the number of missiles deployed. Furthermore, some mobile missiles could be concealed and then deployed during times of crisis to augment the permitted force level. Garrison basing, or deployment on military bases would ease the verification problem somewhat by confining missiles in peacetime to localized areas where they could more easily be counted. If missiles were banned outside these garrisons during peacetime, then the price of covertly deploying them outside these areas would be raised, since any missile that was detected outside the designated areas would constitute a violation.

Ballistic Missile Defence (BMD)

In the course of the debate on SDI many of its supporters have suggested that strategic defences could be deployed to protect US retaliatory forces, and in particular to help to solve the ICBM vulnerability problem. They argue that, while SDI does not include protection of the ICBM forces as a specific goal, protection of the land-based strategic forces would be a natural first step in any BMD deployment. This would both provide extra security for the US retaliatory force and form the foundation for a larger, more capable deployment that could eventually include full protection for the whole population. Furthermore, improving the survivability of the retaliatory forces would minimize the potential instabilities that could arise during the transition to extensive nation-wide defences.

The possibility of active defences raises two questions. Are there combinations of ICBM basing modes and BMD systems that would provide cost-effective responses to future threats? And could such defences be deployed without contravening the current ABM Treaty? The latter question is important not only because of the political implications of a possible US decision to abrogate the Treaty, but also because parallel Soviet BMD deployments could undermine US retaliatory capabilities. Of course, if the Soviet Union decided to break out of the ABM Treaty unilaterally, then defence for retaliatory forces could prove to be an attractive response for the US.

The most important variable in considering future BMD costs and performance is the threat. Any terminal BMD system that must face many thousands of RV (and thousands of decoys as well) is almost guaranteed to fail on grounds of cost-effectiveness. The attacker has the advantages of being able to choose the time of the attack, concentrate it and suppress the defences. It is thus extremely important that the defence should find a way to achieve its objectives without having to defend against the entire threat. In short, the defence must gain leverage by protecting only a few of the aim-points of the attack, and it must defeat defence-suppression tactics by becoming essentially untargetable itself.

The first is possible if the ICBM themselves generate false targets, through mobility or deception, or if the defence is capable of highly effective preferential tactics. The second only comes about through mobility or deceptive basing of the defences themselves, making targeting the defences a less effective tactic than attacking the ICBM sites directly.

Preferential defence is also essential for the effectiveness of any BMD development that stays within the ABM Treaty's numerical limit. For example, 500 small ICBM, in clusters of five scattered randomly over approximately 30,000 square miles, would require an attacker to barrage that area with some 10,000 RV to destroy them. The barrage would put the attacking bursts more than 1½ miles apart, and only one incoming warhead would be close enough to each ICBM cluster to threaten it; thus one interceptor with each cluster (i.e. a total of 100 interceptors) would provide nearly perfect defence. Obviously the defence system would have to be as mobile and as hard as the small missiles, and would have to be undetectable to the attacker before he attacks. Whether, given other options for enhancing ICBM survival, such a BMD system would be cost-effective or a worthy candidate for full-scale development is a separate question. Most studies suggest that, unless the attack size exceeds 10,000 RV, passive defence measures – mobility, deceptive basing, hardening, or some combination of them – are still more attractive.

The case for BMD systems is easiest to make if limits on ABM are relaxed. However, raising the limit on both sides – to 1,000 for example – is scarcely a good option for the US. The extra Soviet defences would kill more retaliatory US warheads than the US defence would save. (This argument assumes that the US would target what the USSR was using its own BMD systems to defend). Naturally, if the Soviet Union abrogates the ABM Treaty first and deploys a defence, or responds to US abandonment of the SALT II numerical limits by greatly increasing its number of RV, US BMD options for protecting ICBM sites might become much more attractive, although leverage is still important so as to keep the system cost-effective.

Terminal BMD systems currently under development – the Exoatmospheric Re-entry vehicle Intercept System (ERIS) and High-altitude Endoatmospheric Defence Interceptor (HEDI), coupled with their associated radars and communications equipment – cannot fully meet most of the above criteria for cost-effectiveness. If limited to 100 interceptors, a defence based on these systems would be easily saturated, as can any fixed system of radars and interceptors at that level. Much larger deployment levels of such systems could effectively protect mobile or deceptively based ICBM sites, provided mid-course tracking systems could predict RV impact points accurately enough to allow some degree of preferential defence, and provided the USSR did not develop effective penetration aids. If the development and procurement costs for these systems were included as part of strategic defence goals other than ICBM protection – for example, as part of the underlay for a general strategic defence deployment – then the marginal costs of ICBM defences would be much lower, improving their cost-effectiveness. However, until strategic defence deployment becomes realistic, defending ICBM sites should be judged without reference to the possibility of more comprehensive defences. From this perspective, it is difficult to justify a terminal active defence for ICBM using the BMD systems currently under development.

Some supporters of SDI would seek to achieve the necessary leverage by using as part of the defence relatively primitive space-based elements, such as chemical rockets ('kinetic kill vehicles') which would attack Soviet ballistic missiles in the boost phase. These would not suffice to provide a 'leak-proof' defence, but they could improve survivability by thinning out the threat and by introducing an element of uncertainty into Soviet attack planning (since the USSR would not be able to assume that any specific warhead would reach its target). However, their deployment would require modification or abrogation of the ABM Treaty and raise issues of cost-effectiveness and possible Soviet countermeasures.

The Outlook

As the debate on the fate of US ICBM proceeds, it will be important to address directly the central strategic question of whether the US needs to maintain a strategic nuclear force posture with survivable ICBM. Decisions should not be taken simply on the basis of the technical arguments over basing and expected costs. Are survivable ICBM necessary to assure a credible retaliatory strategy, given all the other US strategic systems? Without survivable ICBM, will political leaders be prepared to adopt the remaining option – launching vulnerable ICBM on warning or after a few Soviet warheads actually reach US territory?

If the choice is made in favour of survivable ICBM, the United States could proceed in three ways. Unilateral steps could be taken to expand the force, harden it, or make it more mobile; both sides could limit the number of missile warheads through arms control; or defences could be deployed.

Each of these methods, however, has its drawbacks. All the proposed basing modes remain controversial on grounds of technical survivability and political acceptability. Success in arms control depends on the agreement of both sides, which is blocked by fundamental disagreement over the future role of strategic defences. And the deployment of cost-effective missile defences will require modification or abrogation of the ABM Treaty. Moreover, none of these three ways of improving survivability is likely to be sufficient by itself. Unilateral steps will probably need to be combined with constraints on offensive forces or deployment of a defence. Defences will not obviate the need for rebasing or mobile systems, and arms-control limits may not remove the need for defences. Yet, without a change in Soviet attitudes towards defence, only a combination of two of these methods can be pursued simultaneously (rebasing and BMD, or arms control and rebasing). Without a commitment by both sides to forgo defences and maintain the ABM Treaty, it will be difficult to reach an agreement that will achieve significant reductions in offensive weapons systems.

THE DEBATE OVER THE ABM TREATY

Since the launching of the Strategic Defense Initiative (SDI), the Anti-Ballistic Missile (ABM) Treaty has come under increasing scrutiny. Each side has challenged the other's compliance with it, and the desirability of continuing the Treaty regime has become a central issue in both alliance politics and the super-power relationship. In President Reagan's 1983 speech which called for a scientific effort that would render nuclear weapons 'impotent and obsolete', his objective was quite clearly a defence capable of protecting the people of the United States and its allies, and a rejection of deterrence based on the threat of nuclear retaliation. Yet, in signing the ABM Treaty in 1972, the United States and the Soviet Union had each agreed not to deploy ABM systems for the defence of the territory of their own countries and to limit the deployment of ABM to two sites – around a missile field and the national command authority – with a maximum of 200 missiles. (The Treaty was subsequently amended to permit only one site and 100 ABM launchers for each side.) The Treaty's provisions were, however, ambiguous on several counts, and the ambiguities have raised a number of contentious and difficult issues as the US proceeds with the SDI, and as both super-powers deploy new phased-array radars and develop and deploy systems with capabilities similar to ABM: anti-tactical ballistic missiles (ATBM) and anti-satellite (ASAT) weapons.

SDI Research and the ABM Treaty

The US Department of Defense has stated its intention that the SDI programme should be conducted within the provisions of the ABM Treaty. It has defined three basic types of activity that are permitted: conceptual design or laboratory testing (considered during the Treaty negotiations to be research, not amenable to verification by national technical means and not subject to limits); 'field testing' of devices that are not ABM components or prototypes of ABM components; and 'field testing' of fixed land-based ABM components specifically permitted by the Treaty. The bulk of the US short-term effort consists of technology research projects which fall into the first two categories and include the experiments for directed-energy and kinetic-energy weapons. Three experiments involve tests of fixed ground-based ABM components at agreed test ranges.

Nevertheless, some of the planned US SDI activities raise questions as to whether they are in fact permitted under the Treaty and could, over time, have the effect of undermining its provisions, if not actually violating them. The problem is that the various provisions of the Treaty make it difficult to determine precisely what is, or is not, permitted. An example occurred in September 1986, when the US launched a target satellite and used sensors attached to the second stage of the launcher to track and destroy it. Article V of the ABM Treaty forbids tests of space-based ABM components, and Article VI prohibits giving interceptor missiles other than ABM interceptor missiles 'capabilities to counter strategic ballistic missiles or their elements in flight trajectory'. But the Pentagon contended that the test did not fall within the Treaty restraints, because it did not involve an ABM 'component': the satellite travelled too slowly to simulate a real Soviet re-entry vehicle (so this was not a test of an ABM system able to counter an object with the flight trajectory of a ballistic missile), and the intercepting vehicle could not have destroyed the target unless the target had carried a special homing reflector. Critics within the US charged that the test had been conducted at a speed close to that of a forbidden space-based interceptor, and that the sensors used to observe the rocket plume and direct the attack appeared to make it a prohibited space test of a missile interceptor.

Ambiguities in the ABM Treaty include: what will be defined as a 'component' of an ABM system, and how it differs from subcomponents (or what the US terms 'adjuncts'); where development begins and research ends; what capability is needed to counter strategic ballistic missiles; what is meant by testing in an ABM mode; and what is a ground-based (as opposed to space-based) system. Since many more experiments are planned, there will many opportunities to interpret the various provisions of the Treaty, and sharp differences of view can be expected as to whether they are consistent with both its letter and its spirit.

US Reinterpretation of the Treaty

It is clear that the deployment of at least some elements of the strategic defences envisaged under the SDI programme are prohibited by the Treaty. While there are differences between the US and the USSR as to where research ends and development and deployment begin, until October 1985 they both followed the traditional view that the Treaty forbids the development and testing of space-based defences. Under Article V the parties had agreed 'not to develop, test or deploy ABM systems or components which are sea-based, air-based, space-based or mobile land-based'. Both countries acted on the view that this prohibition applied to the technology available in 1972 and to all as yet undeveloped technology. For example, a 1972 State Department report notes that the Treaty precluded the development of systems based upon future technology, and testimony during the US Senate ratification hearings, as well as subsequent reports from the US Department of Defense, reinforce this view. The *Fiscal Year 1985 Arms Control Impact Statement* submitted to Congress by the US Arms Control and Disarmament Agency summed up this policy by noting that 'the ABM Treaty bans the development, testing and deployment of all ABM systems and components that are sea-based, air-based, space-based or mobile land-based . . . The ABM Treaty prohibition on development, testing and deployment of space-based ABM systems, or components for such systems, applies to directed energy technology (or any other technology) used for this purpose'.

In October 1985 the Reagan Administration reversed the previous US interpretation. It declared that the Treaty authorized the development and testing of space-based anti-ballistic missile systems, provided they were based on 'other physical principles' than those employed by the systems in use in 1972. This 'reinterpretation' or 'broad interpretation' focuses on two clauses and one addendum to the Treaty.

Taken alone, Article V would prohibit the development, testing and deployment of a spaced-based defence. Article II(1), however, qualifies this by defining an ABM system as one 'to counter strategic ballistic missiles or their elements in flight trajectory, currently consisting of: (a) ABM interceptor missiles . . . (b) ABM launchers . . . and (c) ABM radars'.

Supporters of the new reading of the Treaty contend that the phrase 'currently consisting of' is a limiting definition, rather than one which is functional or illustrative but not all-inclusive. They also lay particular emphasis on Agreed Statement D, which provides that 'in the event ABM systems based on other physical principles and including components capable of substituting for ABM interceptor missiles, ABM launchers, or ABM radars are created in the future, specific limitations on such systems and their components would be subject to discussion'. The parties, it is argued, would not have adopted an Agreed Statement defining their obligations in the event of their wishing to deploy 'exotic' defences, unless the body of the Treaty failed to regulate these. The State Department's Legal Adviser, in addition, adduces the classified negotiating record to prove that the Treaty does not preclude the development and testing of defences based on future technologies. His claim, that the US delegation tried to secure a ban on these 'exotics' but failed, contradicts the testimony of many of those involved in the negotiations, although the only Administration member involved in the original negotiations supports this view.

In the Administration's view, therefore, the ABM Treaty prohibits only ABM systems based on the technology available in 1972 – which then 'currently consisted of' interceptor missiles, launchers and radars as enumerated in Article II(1). According to this analysis, the Treaty would allow both development and testing of either a space-based particle-beam or a laser strategic defence which are based upon post-1972 technology. It appears that the Administration may also be prepared to argue that the Treaty permits the development of other systems, including space-based kinetic-kill rockets, which would use infra-red sensors (rather than radars) and would be based on 'other physical principles' than those current in 1972.

Opponents of this approach reacted swiftly and vocally when the reinterpretation was announced. One of the Treaty's original drafters spoke for many critics in castigating the Administration for an interpretation that was 'legally, historically and factually wrong', and Gerard Smith, chief US negotiator of the Treaty, stated that this interpretation would make it a 'dead letter'. Such critics contend that the Treaty text, US government pronouncements before October 1985 (particularly those during the Senate ratification procedure) and US and Soviet behaviour under the Treaty since ratification all support the traditional interpretation. They point first to Article I(2), which states as a goal that 'each party undertakes not to deploy ABM systems for a defence of the territory of its country and not to provide a base for such a defence', and contend that this is precisely what the Administration wants its SDI programme to do. They then turn to Article II(1), where an ABM system is defined by reference to its *function* as a system designed 'to counter strategic ballistic missiles or their elements in flight trajectory'. In 1972 these systems consisted of interceptor missiles, launchers and interceptor radars, but this definition, they say, would expand to cover any new technology with the function of countering strategic ballistic missiles. Agreed Statement D was simply a recognition that 'exotic' technologies might necessitate technology-specific constraints, not foreseeable when the Treaty was drafted, in order to maintain the overall treaty regime in the future. This analysis would preclude the development, testing and deployment of any form of defence other than a fixed land-based system.

Statements by government officials at the time of the Treaty's adoption also support this analysis. Secretary of State Rogers stated in 1972 that Article II(1) defined an ABM system functionally. Comments in articles, books and Congressional testimony by those involved in the talks and by other key government figures of the time (such as Secretary of Defense Laird) support the view that the negotiators inserted the word 'currently' to make the definition functional and to close the loophole which allowed the development and testing of 'exotics'.

Members of Congress put special emphasis on the legislative history during the Senate ratification proceedings. They argue that, whatever the negotiators thought they had accomplished, the Treaty, like any other law, must be interpreted according to what the Senate thought it was approving in its deliberations – and this could only be determined from the Congressional hearings and debates before ratification. According to Senator Sam Nunn, any attempt to alter the interpretation of the Treaty which goes against the understanding of the Senate at the time of ratification would raise grave constitutional questions.

Finally, supporters of the traditional view contend that – even if the Treaty were ambiguous, and could plausibly support either interpretation as written – the subsequent behaviour of the parties should determine the boundaries of permissible action. They maintain that, because for thirteen years both super-powers acted on the understanding that the Treaty had forbidden the development of 'exotic' technology defences, the US and the USSR are bound under international law to follow that interpretation until they both agree to modify it.

The Soviet Union also rejected the Reagan Administration's reinterpretation. In an October 1985 *Pravda* article, the Soviet Chief of the General Staff and First Deputy Minister of Defence stated that 'Article V of the Treaty absolutely unambiguously bans the deployment of ABM systems or components of space or mobile ground basing . . . regardless of whether these systems are based on existing or "future" technologies'. A 1986 propaganda brochure echoed this view in stronger terms, characterizing the new interpretation of the ABM Treaty as a 'deliberate fraud'. But the USSR has itself created confusion by insisting that research must be confined to the 'laboratory', without making clear whether or not this is a restatement of the narrow interpretation. Whether it intends to refine this view remains most unclear.

President Reagan sought to defuse criticism of his new policy from Congress, US arms control advocates, the Alliance and the USSR by stating that the US would follow the traditional, or 'restrictive', view as a matter of policy, even though it found the broader interpretation 'fully justified', and he agreed to confine work in the SDI programme to activities consistent with the 'narrow' interpretation. This quelled the furore in Congress for several months, but satisfied neither Alliance members, who continued to push for a return to the traditional interpretation, nor the USSR, which increased its propaganda attacks. And as reports grew that the Administration was preparing to permit SDI experiments which would break the 'narrow' interpretation, members of Congress renewed their demand for Administration adherence to that interpretation. In March 1987 Senator Nunn issued a 98-page report, which he read over three days on the Senate floor, strongly arguing for the 'narrow' interpretation. His speech gave new momentum to Congressional efforts aimed at blocking any Administration attempt to pursue the 'broad' interpretation of the ABM Treaty in conducting tests in the SDI programme.

US and Soviet Compliance with the Treaty

Various Soviet and American activities have also raised a number of issues involving compliance with the ABM Treaty. The USSR's construction of a large phased-array radar near Krasnoyarsk is the most serious; the Reagan Administration has charged that this constitutes a violation because of 'its associated siting, orientation, and capability'. In order to ensure that early-warning radars could not double as ABM radars, the Treaty signatories had agreed not to deploy them except on the periphery of their national territory and oriented outwards. But the Treaty allows such radars to be located anywhere if they are deployed to track objects in outer space, and the

Soviet Union claims that this radar is indeed for space tracking. The consensus in the West, however, is that space tracking is an unlikely primary role, and that the radar violates the Treaty.

The Reagan Administration has reported on four additional compliance issues (mobility of ABM system components, concurrent testing of ABM and air-defence components, SAM system capabilities, and the rapid reloading of ABM), but has been unable to conclude that Soviet activities in this area are Treaty violations. Rather, it has simply claimed that, taken together, they suggest that the USSR may be preparing an ABM defence of its national territory.

US modernization of radar facilities in Thule, Greenland, and the replacement of that at Fylingdales Moor in Britain, have also raised compliance issues. Thule and Fylingdales are far from the 'periphery' of the United States, and so cannot qualify as permitted large phased-array radars; Article IX also forbids either country to deploy radars 'outside its national territory'. The US claims that these radars fall within the exception permitting modernization of facilities that existed at the time of the signing of the Treaty. The Soviet Union claims that they cannot qualify as mere modernizations, or even replacements – because they are a fundamentally different kind of radar (large phased-array), rather than the previous mechanical ones on the sites – and, in the case of Fylingdales, it points to the fact that the new radar is a short distance away from the original site. Whether this modernization programme violates the Treaty remains in dispute. The British government roundly rejects any suggestion that it does so.

In October 1985 the Soviet Union offered to stop work at Krasnoyarsk, if the US would do the same at Thule and forgo the replacement at Fylingdales. The US refused to trade what it characterized as two legal facilities for an unpermitted one. A year later the USSR offered to halt construction at Krasnoyarsk in return for a similar promise from the US over Thule; the US again turned the suggestion down.

ASAT and ATBM

There are two additional US and Soviet weapons programmes implicitly permitted by the Treaty which could weaken the treaty regime, because they involve technologies which could be used for ballistic missile defence.

The first of these, involving anti-satellite weapons, is currently being pursued by the US. The USSR, although it has a crude operational system, claims to be observing the moratorium on ASAT tests. The technologies and components necessary to destroy satellites and ballistic missiles overlap considerably.

The second programme, anti-tactical ballistic missiles (ATBM), poses a particular threat to the Treaty because of the technical similarity between ABM and ATBM – a fact explicitly recognized by the Treaty drafters, who prohibited upgrading of non-ABM missiles to give them the ability to intercept strategic ballistic missiles. The

USSR at present has an advantage in ATBM technology and has deployed systems (SA-10, SA-X-12) with some capability against cruise missiles, intermediate-range ballistic missiles (IRBM) and submarine-launched ballistic missiles (SLBM). Development of systems in this area by either side poses a direct challenge to the Treaty, since some aspects of the theatre threat – notably the Soviet SS-20, or the US *Pershing* II – have characteristics essentially indistinguishable from some 'strategic' missiles, such as SLBM. The possible link between ATBM and ABM has been accentuated by the US decision to award contracts for European theatre ballistic missile defence architecture studies.

The Future of the ABM Treaty

The ABM Treaty is of indefinite duration but is subject to review every five years and to withdrawal after six months' notice. The two signatories reviewed it in 1977 and 1982, and another review is scheduled for the autumn of 1987; whether this will be as *pro forma* as the first two is still unclear.

Decisions on the actual deployment of space-based defences are still some years away. The Reagan Administration has sought Soviet agreement to a co-operative transition, but in the arms-control negotiations it has offered to postpone the deployment of defences bound by the Treaty for ten years, and Defense Secretary Weinberger's efforts to commit the US to an earlier deployment have so far failed. The USSR remains committed, at least rhetorically, to maintaining the Treaty, and this cannot be expected to change in the short term.

Even though neither the Soviet Union nor the United States is likely to withdraw unilaterally from the Treaty, a number of issues concerning its future will need resolution. How each party makes its decision will turn on whether it reckons the balance of national advantage to lie with protecting its own flexibility to proceed with the research and development of missile defences or constraining the other side's potential deployments of them. The two countries have differed in the past and seem unlikely to agree in the future. In the original Treaty negotiations, the USSR sought to retain flexibility and insisted upon ambiguities. This has now become the position of the United States, which has decided not to proceed to tighten up the language, preferring to ensure freedom for the SDI programme even at the expense of not constraining worrying Soviet ABM and ATBM developments.

The United States faces a number of choices. It could seek to modify the Treaty to allow certain types of development and testing not currently permitted under the 'narrow' interpretation. Alternatively, it could seek to strengthen the Treaty by defining its ambiguous terms more tightly. It might have to forgo some scheduled tests, but as part of a compromise the USSR might be required to halt the construction of its early-warning radar and limit its ATBM programme. This option might also permit the US to continue some research and development of a space-based defence, though still prohibiting deployment. The third alternative is simply to leave the Treaty in its present form and try to resolve the existing compliance problems. However, given the technological developments in directed energy, ASAT and ATBM since 1972, this approach would undermine the credibility of the Treaty, perhaps to the point where the US could no longer maintain the Treaty's objectives.

The fundamental issue for the United States is to decide whether to barter constraints on strategic defences for major reductions of offensive weapons in the arms-control negotiations, or whether it will need to deploy missile defences in order to ensure a credible retaliatory strategy. If so, it will have to decide what modifications of the ABM Treaty would be needed and then seek to gain Soviet agreement. Neither option seems likely to be realized in the coming year.

Arms Control

1986 was a year of considerable drama in the arms-control arena, but also one of disappointment. The spirit, if not the substance, of the 1985 Geneva summit meeting persisted in 1986, as General Secretary Gorbachev and President Reagan took an active interest in the full range of US–Soviet arms-control negotiations. Both leaders took personal control of the bilateral arms-control process, advanced dramatic new proposals, and infused the negotiations with an energy and authority that their bureaucrats had previously been unable to provide. Yet no agreements were reached in the centrally important area of strategic nuclear arms, nor yet on intermediate nuclear forces (INF), although here positions converged sharply. Only at the Conference on Disarmament in Europe (CDE) was an agreement reached, when 35 nations adopted new measures to reduce the probability of surprise attack in Europe.

Perhaps most serious was the failure to develop a fundamental framework for a strategic arms-control agreement. Soviet proposals continued to build on the concept of stability established by the SALT process and the ABM Treaty, while US proposals explicitly called for a co-operative transition in the deployment of extensive strategic defences. Moreover, the USSR's activities raised serious questions about its compliance with the SALT and ABM Treaties, and in November 1986 the US exceeded the SALT II Treaty sub-limit on missiles with multiple, independently-targetable re-entry vehicles (MIRV) and bombers carrying air-launched cruise missiles (ALCM). At best, 1986 saw the strategic and arms-control choices facing the negotiators clearly defined.

Bilateral Negotiations

Ever since bilateral arms-control negotiations between the US and the Soviet Union resumed in March 1985, there had been a critical underlying question. Would President Reagan persist with his vision for the Strategic Defense Initiative (SDI), or would he be willing to restrict the programme to research consistent with a strict interpretation of the ABM Treaty in exchange for an agreement on deep cuts in the US and Soviet strategic arsenals? Throughout 1985 the Reagan Administration never had to face this choice directly. The USSR was not willing to allow *any* research on space-based strategic defence – not even research consistent with its own interpretation of the ABM Treaty – and for most of 1985 was unwilling to make any move in the negotiations on strategic and INF systems until the US unilaterally abandoned all space-based defence research.

President Reagan, for his part, had set two preconditions for the pursuit of serious arms-control negotiations: the modernization of

American strategic forces and the initiation of a vigorous SDI programme. These preconditions, together with the diametrically opposed Soviet position on SDI, had stymied negotiations until the November summit meeting in Geneva. That meeting, however, heralded a change in the tenor of the bilateral arms negotiations, as well as the US–Soviet relationship, that raised expectations for 1986.

In 1986 Reagan and Gorbachev began to breathe life into the arms-control negotiations. They opened a direct dialogue, sustained by personal letters and meetings between their foreign secretaries, which continued through the year and culminated in the Reykjavik summit of 11–12 October. This moved the negotiations along at a pace that contrasted sharply with that of the previous year. If anything, it threw up too many proposals; but it invested those proposals with unambiguous authority and engaged the personal and political interests of the two leaders in the process. Most important, it brought the bilateral arms negotiations to a point where President Reagan had to face the critical choice between deep cuts in strategic arsenals and his commitment to deploy strategic defences.

At the Geneva summit, Reagan and Gorbachev had agreed to pursue a 50% reduction in strategic arms – 'appropriately applied' – and an 'interim' INF agreement. On 15 January, however, Gorbachev broke away from this agreement and introduced a proposal to eliminate *all* offensive nuclear weapons by the year 2000. This was divided into three phases: first, both super-powers would over a 5–8 year period implement the Soviet 50% reduction proposal and eliminate their INF in Europe, while Britain and France would freeze their nuclear forces at current levels; then, over 5–8 years starting in 1990, other states possessing nuclear weapons would begin to eliminate their forces; finally, the super-powers would reduce their nuclear forces to zero between 1995 and the year 2000.

Although this proposal was greeted in Washington as a positive development, it had two principal difficulties. The US, as well as France and Britain, had always maintained that British and French nuclear forces could not be the subject of negotiation between the super-powers, and neither Britain nor France are interested in freezing their nuclear forces (which are being modernized), let alone eliminating them entirely. Then, although the US had endorsed the idea of eliminating nuclear weapons in principle, and had proposed a global ban on all INF, Gorbachev's agreement to the elimination of all INF in Europe raised fundamental questions about the credibility of NATO strategy. NATO European governments were therefore unenthusiastic about the proposal and were relieved when it became clear that Gorbachev had put this proposal forward more for its propaganda value than its arms-control value. Even the Soviet negotiators did not seriously pursue the phased elimination of all nuclear weapons at the bargaining table.

The US response, in February, seized upon the one promising element of Gorbachev's 15 January proposal – the elimination of INF in Europe – as a step towards the West's objective of a 'zero option', or a global ban on INF. Moreover, by allowing French and British forces to remain intact in the first phase, Gorbachev's proposal seemed to be dropping the previous Soviet requirement that the USSR should be compensated for the existence of French and British nuclear forces in any INF agreement. The US response called for global elimination of all INF within three years. This represented a shift from the more modest US proposal in the autumn of 1985, which had called for global equality in INF warheads, an interim limit in Europe of 140 launchers each for the US and the Soviet Union, and proportional reductions of Soviet launchers in Asia. In addition, Reagan reiterated his interest in eventually eliminating nuclear weapons, the importance of achieving 50% reductions in strategic offensive weapons, and the urgency of redressing the conventional imbalance in Europe.

The speed with which the US counter-proposal was put together, and the fact that it was delivered directly to Gorbachev as well as being publicized in a statement, reflected the new character of the bilateral arms-control dialogue: it was being conducted both in public and at the highest level. This resulted in two, somewhat contradictory, trends. First, in an effort to mobilize public opinion behind their proposals, the two leaders (particularly Gorbachev) articulated broad, publicly appealing arms-control objectives which generated public interest in the arms-control process but were inherently difficult to negotiate. Second, the public interest that had been aroused put each leader under pressure to respond quickly with proposals that seemed to be moving the process forward. As a result the arms-control dialogue began to develop proposals that were based on radical objectives, but at the same time contained elements of flexibility regarding the core issues.

The Soviet proposal of 15 January had put the US on the defensive, but Reagan's swift and public response neutralized much of the impact of Gorbachev's call for global disarmament. In addition, Gorbachev's public diplomacy was not assisted by the delay in the Soviet announcement of the meltdown of the Chernobyl reactor on 25 April. This put the USSR on the defensive for the first time in 1986, and its subsequent arms proposal (tabled in Geneva on 11 June) contained true indications of flexibility. Consistent with the discontinuity in the bilateral arms-control negotiations since the Geneva summit, this new proposal turned from INF to the issues of strategic offensive arms and space and defence. The USSR had previously insisted that all 'space strike weapons' – and the research, development and testing of them – should be banned in exchange for a 50% cut in strategic weapons (defined as weapons that could strike the other side, thus including US forward-based systems). The new Soviet initiative proposed to allow research, development and testing of SDI components, as long as these activities were confined to the laboratory – provided the US would agree not to withdraw from the

ABM Treaty for 15–20 years. In addition, it was proposed that these agreements be complemented by a more modest 30% cut in strategic weapons, rather than the 50% agreed at the Geneva summit.

This new position represented a critical turning point in the negotiations. Not only did the USSR concede that some types of SDI research could be allowed, but the structure of this proposal – a strengthened ABM Treaty in return for cuts in strategic weapons – was to remain the basis of its negotiating position through the Reykjavik meeting. There were still fundamental problems with the Soviet offer – such as the definition of 'strategic systems' and 'laboratories' – but the proposal provided a basis for true negotiation. The previous Soviet position on space-based defence research had been totally irreconcilable with the SDI programme. Now, although a great gap remained between the US and Soviet positions, in principle some balance might be found between the US desire for research on space-based defences and the Soviet desire to prevent the US developing and deploying 'space strike' weapons. In addition, on 23 June Gorbachev sent Reagan a letter which stated his willingness to compromise on INF, although he offered no specific proposals. The modest goals set by the Geneva summit had been expanded as significant achievements beckoned, and progress was being made in all three areas of negotiation.

The US response came in a letter from Reagan to Gorbachev on 25 July. Not only did its proposals represent a radical departure from previous US positions, it had been formulated by a small group of top-level advisers supervised by the President, rather than emerging from the normal working groups made up of representatives from the Departments of State and Defense and the National Security Council. This 'top-down' approach was to characterize the White House's approach to arms control for the next three months.

Reagan's letter covered the entire gamut of issues relevant to US–Soviet relations. Much of it did not break new ground: it reiterated the 22 February proposal on INF, and it repeated the current US offer of a 50% cut in strategic weapons (though allowing that this could be staged: an interim cut of 40%, followed by discussions about the other 10%). Its critical element was a three-part proposal offering compliance with the ABM Treaty for at least 7½ years. Until 1991 both the US and the USSR would agree to confine themselves to research, development and testing consistent with the ABM Treaty, in order to assess the feasibility of strategic defences. A new treaty would be signed immediately stipulating that, if either side decided to deploy strategic defences after 1991, then it would be obliged to offer a plan, to be negotiated within two years, for sharing the 'benefits of strategic defense' and eliminating offensive ballistic missiles. However, if the two sides could not agree on such a plan after two years of negotiation, then either side would be free to deploy advanced strategic defences after giving six months' notice.

This proposal presented difficulties for the USSR. It was objectionable because it did not really constrain SDI. Not only did it leave open the question of whether the US would use a broad or narrow interpretation of the ABM Treaty, it gave the US the right eventually to deploy strategic defences without the USSR having any influence over that decision. As a White House official commented: 'The President's intention is to get the Soviets to agree to deployment at the end of seven years in return for an American commitment not to withdraw before then'. Since deployment of strategic defences could not take place for at least seven years in any case, even by optimistic estimates, this offer had limited attraction for the USSR. Moreover, any US–Soviet collaboration over deployment of strategic defences was linked to an attempt to eliminate offensive ballistic missiles. Eliminating these from the Soviet armoury would leave only bombers and cruise missiles: two types of strategic weapon in which the US has a considerable advantage. Thus, the specifics of the 25 July proposal provided no grounds for agreement, although some elements, both in themselves and in their relationship to one another, were roughly analogous to the USSR's June offer and provided a real basis for negotiation. In fact, American officials had expected not agreement but a Soviet counter-proposal that would be the subject of discussion in the next round of Geneva talks and at any future summit.

Some US officials had doubts about the proposal, too. The Joint Chiefs of Staff had reviewed the draft and were troubled by the proposal calling for the elimination of offensive ballistic missiles, particularly by its implications for NATO strategy and extended deterrence. However, because they considered the negotiating process to be in a nascent stage – still sufficiently conceptual for them to endorse the proposal as a broad goal 'in the abstract' – they neither analysed the military implications of a ballistic missile ban nor tried to reconcile the substance of the US proposal with US and NATO strategic policy. A gap between strategy and arms-control policy had previously opened up with the INF 'zero option' proposal, and another was developing in the strategic weapons area. These gaps would become increasingly important as the negotiating process moved from rough agreement on a conceptual approach to a stage in which the negotiators focused on specific proposals, and bargaining could no longer be conducted 'in the abstract'.

The Road to Reykjavik

During the run-up to the sixth round of Geneva talks, due to open on 18 September, three possibilities for the negotiations were emerging. President Reagan could remain committed to his broad vision for SDI, not tolerating any constraints on the research or deployment of strategic defences: in that case the negotiations would deadlock over fundamentals, and no agreement would be reached. He could make the 'grand compromise' that many hoped he might be planning: using SDI as a bargaining chip, he could agree to 50% reductions in strategic offensive weapons and renew the US commitment to the ABM Treaty. Or he could seek modest reductions (for example, 30%) in strategic offensive weapons in exchange for an agreement on SDI that would finesse US–Soviet differences over the future of strategic defences: a mutually acceptable definition of research might be reached, and a commitment made on the basis of it to remain in compliance with the ABM Treaty for ten years, but no commitments would be established beyond ten years. In each case an INF agreement would be pursued independently.

To achieve the second or third possibilities, as both sides seemed intent on doing, progress in developing a mutually acceptable conceptual approach to an arms-control agreement would have to be translated into progress on substance: the specifics of proposals. This would be necessary not just to sustain any chance of agreement, but also to provide the two leaders with specific points of discussion if there should be a summit meeting in 1986, as had been agreed at the 1985 Geneva summit. In addition, those involved in the negotiating process were beginning to need to keep an eye on the US political calendar if hopes for an agreement during the Reagan presidency were to be fulfilled. It appeared that the framework of an agreement would have to be established by December 1986, leaving nine months to finalize negotiations in preparation for a final summit in the autumn of 1987, if there were to be time for Congress to ratify a treaty before the 1988 presidential election campaign made this impossible. Perhaps for these reasons, as well as some of its own, the USSR expressed a desire to continue bilateral discussions before reconvening in Geneva in mid-September. The United States agreed.

The two rounds of bilateral discussions – in Moscow in August and Washington in September – brought no substantial changes in negotiating positions. Rather they served to explore some of the larger questions underpinning the US and Soviet proposals, and in doing so reflected both sides' desire to work around the fundamental differences between them. The USSR showed considerable flexibility in its INF position, expressing interest in an interim agreement on INF that would set each side a limit of 100 warheads in Europe, and confirming that the issue of British and French forces would not be an obstacle. It also suggested that there could be some reductions in Asia, although it would not commit itself to the elimination of INF which the US sought.

At the opening of the September round in Geneva the US responded by modifying its INF and START proposals. These modifications, along with Reagan's 25 July letter, formed the basis of the American position going into the October meeting in Reykjavik (see Box 1 on next page).

1: US PROPOSALS

Strategic Offensive Arms Reductions
- 1,250 ICBM, SLBM
- 350 heavy bombers (no limit on gravity bombs and short-range attack missiles (SRAM))
- 5,500 ballistic missile warheads
- 2,000 air-launched cruise missiles (ALCM)
- ICBM limited to 3,300 warheads, with no more than half on heavy missiles, on missiles with more than six warheads, and on mobile missiles
- 50% reduction in Soviet throw-weight
- ban on new, heavy ICBM
- ban on modernization or replacement of heavy ICBM
- ban on mobile ICBM unless the USSR could convince the US that numbers could be verified

INF (new proposal introduced 18 September)
- global limit of 200 warheads for each side
- sub-ceiling of 100 warheads in range of Europe
- collateral constraints on shorter-range missiles (i.e., SS-12 (mod.), SS-23)
- effective verification, to include on-site inspection as necessary

Space and Defence
- 3 stages:
  1) research, development and testing consistent with the ABM Treaty through 1991
  2) treaty stipulating that if either side wants to deploy strategic defences after 1991, that side must present a plan (to be negotiated over a two-year period) for 'sharing the benefits of strategic defence' and for eliminating offensive ballistic missiles
  3) if no agreement on plan, either side allowed to deploy strategic defences after 6 months' notice

The USSR also refined its proposals in September in a letter from Gorbachev to Reagan. These refinements, in conjunction with previous proposals, constituted the Soviet position going in to Reykjavik (see Box 2 opposite).

In addition to offering amendments to Soviet proposals, Gorbachev included in his September letter to Reagan an invitation to a 'private working meeting' in either Reykjavik or London. Throughout the year the question of whether Reagan and Gorbachev would meet in the US, as they had agreed to do at the Geneva summit, had been a source of friction in US–Soviet relations. Gorbachev had insisted that the next meeting between the two leaders should be one of substance, specifically with regard to arms control; Reagan, on the other hand, had been eager that a summit should take place in the US regardless of progress in the arms-control negotiations, and that any meeting should cover the range of issues relevant to US–Soviet relations. Gorbachev's offer of a working meeting to prepare for a summit provided an opportunity to accommodate both leaders' interests. A full-scale summit might take place later in the year, which addressed Reagan's concern; meanwhile a working meeting on arms control would give Gorbachev the chance to test the US position for any flexibility that might lead to progress on strategic defence. Moreover, US–Soviet relations had deteriorated drastically over the arrest of a Soviet spy, Gennady Zhakarov, in the US and the counter-arrest of *US News & World Report* journalist Nicholas Daniloff in Moscow. That problem was still unresolved, and the two leaders, particularly Gorbachev, did not want the affair to sour the progress that had been made on arms control. To maintain the negotiating momentum, the US and the Soviet Union announced on 30 September that both men were to be returned to their respective countries and that Reagan and Gorbachev would meet in Reykjavik.

2: SOVIET PROPOSALS

Strategic Offensive Arms Reductions (strategic defined as capable of reaching the other side's territory)
- 1,600 ICBM, SLBM, heavy bombers
- 8,000 'nuclear charges', including gravity bombs, SRAM and cruise missiles with ranges greater than 600 km.
- no more than 60% of the total 'nuclear charges' could be deployed on any one leg of the triad
- ban on sea-launched cruise missiles (SLCM) on surface ships; limits on SLCM on nuclear submarines
- ban on new types of strategic delivery systems
- verification measures beyond national technical means, including restricting deployment areas for mobile missiles
- agreement on strategic offensive reductions linked to US agreement on defence and space

INF
- interim agreement to limit warheads in Europe to 100
- US to eliminate *Pershing* II missiles
- verification to include on-site inspections as necessary
- British and French forces not included in any bilateral agreement
- perhaps some reduction of INF in Asia

Defensive and Space Weapons
- mutual pledge not to withdraw from ABM Treaty for 15 years
- limit work on 'space strike weapons' to laboratory research
- stop work on Krasnoyarsk radar if US does the same with two radars in England and Greenland
- US adherence to current Soviet ASAT testing moratorium
- 'necessary' verification

At the Summit

The US arrived in Reykjavik expecting, in Reagan's words, 'essentially a private meeting' between the two leaders which would not result in any 'substantive agreements'. Indeed, on 7 October the Administration had announced that it would not present any new proposals at Reykjavik. In contrast Gorbachev arrived ready to deal. At the first meeting, on the morning of Saturday 11 October, Gorbachev surprised Reagan by producing a stack of prepared papers from which to work. Reagan talked broadly about his goals for the SDI, his idea of linking deployment of strategic defences to the elimination of ballistic missiles, and his willingness to share SDI benefits. Gorbachev made specific proposals which included a 50% cut in strategic nuclear weapons, complete elimination of INF in Europe, no withdrawal from the ABM Treaty for ten years, and a strengthening of ABM Treaty language on testing (i.e., restricting research and testing of space elements to the laboratory).

The proposals for a 50% cut in strategic nuclear weapons and a strengthening of ABM Treaty language were not new; but those to preclude withdrawal from the ABM Treaty for only ten years and to eliminate all INF in Europe went beyond what the USSR had presented in Geneva. The entire statement reflected Gorbachev's intention to achieve some breakthroughs in Reykjavik, or at least severely test Reagan's flexibility. During the discussions that followed this initial exchange, Gorbachev rejected the offer to share SDI technology and expressed dissatisfaction with Reagan's explanation of SDI. Reagan rejected the Soviet offer to eliminate INF in Europe without also reducing Soviet INF in Asia. The first session concluded without much progress, but a bargaining process, rather than simply an exchange of views, had begun.

In the afternoon session the US proposed that US/Soviet working groups should meet during the night to flesh out the areas of agreement and disagreement between the two sides. Gorbachev agreed, and both leaders determined that the goal of the groups would be to reach an agreement on instructions to foreign ministers, who would then work towards a 'framework agreement' for signature at a Washington summit. Thus a meeting that the US had expected to be an exchange of views was becoming a full-scale negotiating session.

Compromise . . .

Reagan and Gorbachev had sanctioned two working groups. One would discuss bilateral, regional and human-rights issues. The other, led by special Adviser on Arms Control Paul Nitze for the US and Marshal Sergei Akhromeyev for the Soviet Union, would work out the differences on arms control. During the arms-control working group session – which lasted all Saturday night – the USSR made additional concessions in START and INF.

Since Gorbachev had agreed to eliminate all INF in Europe, the two principal outstanding issues were the Soviet INF in Asia and the short-range missiles in Europe. In the working group meeting the USSR agreed to freeze its short-range missiles, both in Europe and in Asia, and at the start of the two leaders' Sunday morning meeting Gorbachev went further and agreed to reduce all Soviet INF warheads outside Europe to 100. This was compatible with the US 18 September proposal that the USSR should be limited to no more than 100 INF warheads in Asia, and that the US should be allowed to retain 100 INF warheads on US territory. In addition, Gorbachev said that he would be willing to proceed with negotiations on reductions in short-range missiles.

In sum, the Soviet concessions on INF had eliminated the last of the major differences that had existed between the US and Soviet positions at the beginning of 1985. Moreover, during the all-night working session the Soviet Union had agreed in principle (it wanted the details to be worked out at Geneva) to three US verification requirements: a data exchange on numbers before and after reductions; on-site monitoring of missile destruction; and monitoring of the missile factories. Thus, the major elements of a draft agreement on INF would be:

- elimination of all INF in Europe
- a global limit for each side of 100 INF warheads outside Europe
- a freeze on short-range missiles, and subsequent negotiations on reductions and residual levels
- verification (to include the three principles described above).

The only major departure from the US 18 September INF proposal was the elimination of all INF in Europe. There was no consultation with the NATO allies before Reagan agreed to this step on Sunday morning. Immediately after the offer, however, the President authorized telephone calls to the leaders of the European countries committed to INF deployment, informing them of the US intention to agree to the elimination of INF in Europe.

Because the USSR had always held the agreed goal of 50% reductions in strategic offensive weapons hostage to an agreement on strategic defence, the actual implementation of 50% reductions had never been the focus of sustained serious negotiation. During the Saturday night working group session in Reykjavik the US and the USSR attacked two issues that had plagued strategic offensive arms reduction talks for years: what weapons should be defined as strategic, and how they should be counted under the overall ceiling.

As an introduction to the discussion on offensive strategic weapons the Soviet negotiators produced a chart of proposed reductions. This was most notable for what it did not contain. There was no reference to forward-based US systems that could reach the Soviet Union, which had previously been part of a Soviet definition of strategic weapons liable to reduction. Over and above the changed definition – which was a major, though expected, step forward – the USSR also agreed to a reduction of strategic nuclear delivery vehicles (SNDV:

ICBM, SLBM and heavy bombers) to 1,600 on each side, a limit of 6,000 offensive strategic warheads, and reductions in Soviet heavy ICBM (which had been of particular concern to US strategists since SALT I). It also agreed that separate sub-limits would apply to ICBM and that SLCM (which had been a source of irreconcilable differences) should be discussed in future negotiations.

Finally, towards the end of the all-night working group session the USSR proposed new bomber counting rules within the framework of the 50% reductions already agreed. Previously it had insisted that each weapon carried aboard a strategic bomber should be counted as a single 'nuclear charge'. This meant that a simple gravity bomb would be reckoned as the equivalent of an ICBM warhead. Yet a gravity bomb is inherently less capable than an ICBM warhead – it takes longer to arrive at its target and is delivered by a system that is vulnerable to Soviet defences. The US position had therefore been that only ALCM, and not gravity bombs and SRAM (which are used for air defence suppression), should be counted towards a warhead limit. Under the new rules now proposed by the USSR, each ALCM aboard a strategic bomber equipped only with ALCM would count towards the 6,000-warhead ceiling; a bomber with only gravity bombs or SRAM would count as one warhead towards the 6,000-warhead ceiling, whatever the actual number of these weapons it was carrying. Whether the number of ALCM would be counted as the 'maximum equipped' designation in SALT II or the number actually deployed, and how bombers equipped with both ALCM and gravity bombs would be counted, was left unclear. Nevertheless the US immediately accepted this counting rule.

Reykjavik Proposals and 1986 Inventories Compared[1]

		1986	
	Reykjavik	US	USSR
Strategic Nuclear Delivery Vehicles	1,600	1,910	2,502
Nuclear Warheads[2]	6,000	11,306	10,556
INF warheads - Europe	0	316[3]	922
INF warheads - Global	100	358[4]	1,435

[1] For US–Soviet nuclear weapons as of 1 July 1986, see *The Military Balance 1986–1987* (IISS, London: 1986). For an analysis of how to estimate the US–Soviet nuclear balance see pp. 218–22.
[2] Warhead totals were derived using the following bomber counting rules: for bombers with ALCM, each counts as 20 warheads (as in SALT II); for bombers with gravity bombs/SRAM, each counts as one warhead; for bombers with ALCM and gravity bombs/SRAM, only the ALCM are counted.
[3] This includes the additional 80 GLCM the US has deployed since July 1986.
[4] This figure includes a training battalion and 42 *Pershing* II launchers the US has deployed in the US as possible replacement or reinforcement systems.

These developments in the US and Soviet positions on strategic offensive reductions represented significant, but not conclusive, progress. Unlike the final position on INF, many differences remained: on including sub-limits for ICBM warheads and throw-weight; on modernization and replacement of delivery systems; and on verification of mobile missiles. Nonetheless, the major structural differences between the US and Soviet positions had been erased, and the framework for an agreement on reductions which would form a good basis for further negotiations in Geneva had been achieved. (See Table for a comparison of this framework and the 1986 levels of US and Soviet strategic nuclear forces and INF.)

The progress on INF and strategic offensive systems made in the first twenty-four hours of the Reykjavik meeting surpassed all that had been achieved at Geneva during the previous twenty months. An American characterized the process as 'stunning', and a Soviet official called it a 'whirlwind'. Throughout the discussions, however, the issue of strategic defence consistently remained a stumbling block.

. . . and Confusion

When the current bilateral arms-control discussions began in Geneva in March 1985 the Soviet Union had insisted that progress on INF and strategic offensive systems was contingent upon the resolution of differences over strategic defence. Subsequently INF had been unlinked, but an agreement on strategic defences was still a prerequisite for progress on reductions in strategic offensive weapons. Throughout the Reykjavik discussions the relationship between the three areas remained unclear. Two critical questions were outstanding. Would Reagan strike the grand compromise and accept constraints on SDI in exchange for deep reductions in offensive arms? And would Gorbachev consummate the progress already made at Reykjavik unless there were a break in the deadlock over space-based defences?

During the Sunday morning discussions the USSR reiterated its position on strategic defences: reductions in strategic offensive weapons were contingent on a ban on the deployment of space-strike weapons and severe constraints on relevant research and testing. But the agreements reached on strategic offences during the all-night session created a new context for the examination of strategic defences. To explore this, Gorbachev suggested, and Reagan agreed to, an unscheduled Sunday afternoon session to discuss strategic defences further.

During lunch Reagan agreed with his aides to accept Gorbachev's proposal that both the US and the USSR should pledge not to withdraw from the ABM Treaty for a period of ten years. In addition, his closest advisers convinced him that a proposal to eliminate all ballistic missiles over the same period should be added. This idea had been mentioned in Reagan's 25 July letter to Gorbachev, though without a time frame. The main elements of the new US position were:

- both sides should pledge not to withdraw from the ABM Treaty for ten years

- adherence to this pledge for the first five years would be contingent upon the implementation of a 50% cut in strategic offensive weapons
- adherence to the pledge for the second five years would be contingent upon the elimination of all ballistic missiles during that period.

Although it adopted the Soviet ten-year pledge in name, the US position was still objectionable to the USSR for two principal reasons. First, it did nothing to address Soviet concerns about the SDI: the US retained the right to interpret broadly the ABM Treaty restrictions on research, development and testing of ballistic missile defences; ten-year adherence to a flexible interpretation of the Treaty would not significantly constrain SDI; and both sides would be free to deploy after ten years. Second, an agreement to eliminate all ballistic missiles would be distinctly disadvantageous to the USSR, because the Soviet nuclear deterrent consists mainly of ICBM, and the US bomber and cruise missile forces are quantitatively and qualitatively superior (particularly with the development of 'stealth' technology for bombers). The proposal was consistent with Reagan's vision of a world in which strategic defences replaced ballistic missiles as a deterrent, but the USSR regarded this as completely unacceptable, and its response offered no hope of a compromise. Gorbachev repeated his Saturday morning offer and added the stipulation that the eventual deployment of strategic defence would be contingent on future negotiations. Perhaps because he was unwilling to reject explicitly a proposal to eliminate ballistic missiles, he agreed to reduce strategic offensive weapons by 50% in the first five years, but also called for the elimination of all remaining strategic nuclear weapons (not just ballistic missiles) during the second five years. Thus the main elements of the Soviet response were:

- both sides should pledge to comply with the ABM Treaty for ten years
- the Treaty should be strengthened, so that testing of 'all space elements of ABM defence in space are prohibited, except research and testing in laboratories'
- the eventual deployment of strategic defences should be contingent on future negotiations
- a 50% reduction in strategic offensive weapons during the first five years (by 1991)
- elimination of all remaining strategic nuclear weapons during the second five years (by 1996).

This counter-offer showed little flexibility on the question of strategic defences. Many aspects – such as the definitions of 'research', 'testing' and 'laboratories' – were still unclear, but instead of focusing on them, or else putting them aside for further negotiation in combination with the agreed principles for reductions in offensive forces, Gorbachev and Reagan became preoccupied with proposals for general disarmament and visions of perfect defences. Reagan reasserted that both sides should eliminate offensive ballistic missiles over a ten-year period, and should then be free to deploy strategic defences. Gorbachev responded by calling for the elimination not just of strategic nuclear weapons, but of all nuclear weapons. It is not clear whether Reagan gave verbal agreement to this suggestion. Whatever the case, the negotiations had deteriorated into vague and unrealistic proposals by both leaders. No common ground could be found between the two sides' positions on strategic defence, and the meeting foundered, with some bitterness, over this failure. Moreover, Gorbachev refused to sanction any of the other agreements – including INF – until the strategic defence issue was resolved.

Bilateral Negotiations Since Reykjavik

The Reykjavik meeting settled many of the bilateral arms-control issues that had eluded negotiators for years. An INF agreement had been all but signed, and the basic framework for 50% reductions in strategic offensive weapons had been established. But during the meeting discord over strategic defences prevented any agreement. There was considerable discussion of strategic defences, and the areas of disagreement were clarified, but the US and the USSR did not really explore possible areas of compromise on research, development and testing.

This failure was all the more disappointing in contrast to the remarkable breakthroughs achieved in the other areas. However, unlike the long-standing discussions on INF and strategic offensive weapons, serious dialogue on strategic defence had only just begun. The issues raised were complex – requiring careful review, followed by extensive discussion in a negotiating forum such as Geneva – and were not amenable to resolution in the few hours Gorbachev and Reagan devoted to them in Reykjavik. The two leaders tried to force through their differing positions on the basis of inadequate consultation, instead of simply agreeing to acknowledge the successes achieved at Reykjavik and returning the issues to their Geneva negotiators. The result was that they ended up setting the negotiations on a course towards fundamental disagreement, rather than either the 'grand compromise' or modest agreement that had been real possibilities. That course was not irreversible, but it would be difficult to reverse.

The next high-level encounter between the US and the USSR took place in Vienna on 5–6 November, when Shultz and Shevardnadze met at the convening of the Conference on Security and Co-operation in Europe (CSCE). Those who had been disappointed with the final outcome of the Reykjavik meeting looked for a resumption of the progress and momentum achieved there. However, US–Soviet acrimony over diplomatic expulsions and counter-expulsions (catalysed by a US demand in September that 25 Soviet diplomats should leave the United Nations in New York) and the complexity of the issues left on the table precluded this.

On 27 October President Reagan signed National Security Decision Directive 250, which established the final US position in Reykjavik as the new baseline for arms control and provided the basis for a new set of instructions for the US negotiators, who returned to Geneva in January 1987. Soviet negotiators also returned to Geneva ready to pick up where the Reykjavik meeting left off. Working groups were established in each of the three areas to continue to resolve differences and draft the language for possible agreements.

Then, in February, Gorbachev took the initiative and called for the elimination of INF missiles in Europe, with a limit for each side of 100 warheads elsewhere (the Reykjavik formula), and agreed to proceed without first resolving differences over SDI. The US responded by introducing a draft position which agreed in principle but called on the USSR to station its 100 warheads out of range of Western Europe and Japan. The only remaining issues in dispute are where the two sides would be allowed to deploy their 100 warheads, what constraints should be put on shorter-range missiles and the details of verification.

In START both sides have adhered to the framework agreed in Reykjavik. The remaining dispute centres on sub-limits within the 6,000-warhead total. There is some disagreement as to whether the USSR accepted any specific constraints on its heavy missile warheads at Reykjavik, but, apart from verification provisions, the sub-limit question remains the only substantial obstacle to agreement in principle. But here the Soviet position is still linked to agreement on strategic defence. A working group has been established to resolve differences over the interpretation of the ABM Treaty, and dealing with these could be the key to convergence between the two positions. Current indications, however, are not favourable, and the US may soon adopt a broad interpretation of the ABM Treaty. Moreover, Reagan is reviewing proposals for an early SDI deployment that would contravene the ABM Treaty. The Administration will be consulting with its allies and Congress before any final decisions are made, though, and it will come under heavy pressure to remain in strict compliance with the Treaty.

Implications for NATO Strategy

The Reykjavik meeting dramatically and starkly revealed the extent to which Western arms-control policies have diverged from NATO strategy. Since the onset of approximate strategic nuclear parity in the early 1970s there has been growing unease about NATO's strategy of Flexible Response, which must credibly threaten the first use of nuclear weapons in order to offset Warsaw Pact conventional advantages. The arms-control process had contributed to this unease, in that the SALT regime sought to institutionalize nuclear parity and achieve strategic stability through a negotiated balance in which the basic role of nuclear weapons is to deter their own use. Although this objective was difficult to reconcile with the fundamental assumptions of Extended Deterrence, the SALT process had offsetting benefits, primarily its contribution to relaxing tensions between East and West.

NATO's December 1979 decision to deploy US INF on European soil was an attempt to restore credibility to the doctrine of Extended Deterrence. Opinions vary as to whether credibility is enhanced by the political symbolism of the deployment, by its diversification of NATO nuclear options to match those open to the Soviet Union, or both. The deployment was criticized on the grounds that there was no unambiguous military or political rationale for it, but in fact the ambiguity was what made the deployment effective. Extended Deterrence in NATO is no longer based on concepts such as escalation dominance but on cultivating doubt in Soviet minds about what the consequences of conflict in Europe might be. To the extent that NATO could increase such doubt by deploying a range of credible nuclear options, the credibility of Extended Deterrence was enhanced.

But West European public opinion was doubtful about the need for the new missiles, and a tension developed between the declared military necessity for them and domestic political willingness to see them installed. NATO sought to reduce this tension by committing itself to arms-control negotiations to remove the need for INF even as its governments fought to deploy them. Moreover, it then (in 1981) adopted a negotiating position that was politically attractive, yet unlikely to be accepted: the zero option. This option – which Moscow promptly rejected – both dampened public criticism of the deployment and also addressed NATO's military problems, since it seemed unlikely that arms control would in the event hinder the deployment. The long-term risk that the zero option might eventually be accepted did not seem great enough to preclude its use for short-term gain.

The US INF were gradually deployed in Europe from 1983 onwards. The problems of the conventional imbalance and the credibility of Extended Deterrence remained, but they were simmering quietly in NATO committees reviewing the issues. The Reykjavik meeting unexpectedly brought them to the boil. Hitherto the Alliance had managed to paper over its strategic problems by maintaining both its nuclear options and the impression that any general conflict in Europe would escalate to the use of nuclear weapons. But Reykjavik threatened to undermine this. Agreement was reached – subject to acceptable understandings on SDI – that INF in Europe should be eliminated. In addition, aspects of a reduction in strategic offensive weapons to 6,000 warheads were discussed, and, much to his allies' surprise, Reagan proposed the elimination of ballistic missiles over the next ten years. Not only would such proposals call into question the credibility of the US nuclear umbrella, but the significant, and so far intractable, conventional imbalance would then become more dangerous.

The Reykjavik meeting served to concentrate minds on the viability of NATO strategy. While achievement of the broader goals of disarmament and the total elimination of ballistic missiles is not likely, the removal of INF from Europe and the initiation of follow-on negotiations to constrain shorter-range missiles have been made more likely by Gorbachev's February 1987 decision to break the link between agreement on INF from agreement on strategic defences. This transfers much of the burden of ensuring a credible NATO deterrent onto conventional forces, and thus has implications both for national force levels and for the objectives to be pursued in the multilateral conventional arms-control negotiations.

In this sense the Reykjavik meeting also served to make explicit the link between nuclear arms control and conventional arms control. This link was unambiguously acknowledged by NATO governments, as well as by the NATO military command (which felt that it had not been sufficiently consulted), in the aftermath of the Reykjavik meeting. For example, West German Chancellor Kohl, meeting with Reagan at the end of October, emphasized his concern about Soviet conventional force superiority, and stressed that 'NATO's strategy would be deprived of its credibility' if reductions such as those envisaged at Reykjavik were to be made in isolation from necessary adjustments in conventional and chemical forces. Unfortunately, despite success in the CDE, multilateral negotiations concerning chemical and conventional forces in 1986 did not achieve the progress necessary to effect the required adjustments.

Multilateral Issues

Multilateral arms-control negotiations had a mixed record of success in 1986, but they nevertheless showed a degree of innovation. The most significant example was the CDE in Stockholm. Commissioned by the 1983 Conference on Security and Co-operation in Europe (CSCE) Review Meeting in Madrid, it convened in January 1984, with all 35 CSCE nations participating, and concluded in September 1986 with a document on confidence- and security-building measures (CSBM). The goal of the CDE was to build confidence by reducing the secrecy surrounding military activities in Europe 'from the Atlantic to the Urals', and thereby to help to reduce the possibility of war through either surprise or miscalculation. It therefore involved both an effort to build on the confidence-building measures (CBM) in the Helsinki Final Act of 1975 and an operational approach to arms control, focusing on guidelines for the conduct of military activities (as opposed to the structural approach of MBFR, which seeks to reduce military forces in central Europe).

The central provisions of the agreement reached in Stockholm require an exchange of calendars of significant future military activities; notification 42 days in advance of military activities involving more than 13,000 troops or 300 tanks; observation of military activities involving 17,000 or more troops; and ground and aerial inspection to verify compliance. (For the provisions, see *Survival*, January/February 1987, pp. 79–84.) By their nature, these provisions primarily provide information and a limited capability to verify its accuracy, thus complementing intelligence derived through 'national technical means' (NTM). In the end, though, the CDE failed to mandate a detailed exchange of information on the military forces in Europe, and its verification procedures are restricted. Each country can only be inspected three times per year, and aerial inspections are to be carried out by means of aircraft and pilots provided by the host country. However, a modest constraint measure was negotiated, requiring a notifiable military activity in excess of 75,000 troops to be placed on the calendar two years in advance.

Although limited, the CDE agreement does move towards reducing the secrecy of military operations. Its major achievement is symbolic, but still politically important. Unlike the 1975 Helsinki Accords, its provisions are mandatory, the verification procedures go beyond simple agreement in principle to arrangements for on-site inspection, and the inspection procedures are spelled out. Most important, East and West were able to achieve an arms-control agreement in a negotiation that required considerable compromise and hard bargaining.

The provisions in the Stockholm Document only represent commitments to provide information and do not change patterns of military activity, so they do little actually to reduce the prospect of war. If the follow-on negotiations in CDE are to promote fully the objective of preventing a surprise attack in Europe, they will need to go beyond the Stockholm provisions to expand the definition of military activities subject to notification and observation, achieve a detailed exchange of information on the structure of ground and air forces in Europe, improve verification provisions, and constrain threatening military activities. The dilemma for the West is that in seeking to constrain the East's military activities it must accept less flexibility and freedom in its own. This could be particularly dangerous if it degrades NATO's ability to mobilize its reserves and call on US reinforcements in a crisis.

The MBFR Talks

The operational approach to arms control embodied by the Stockholm meeting could be carried further, but there is a limit to how much such an approach can contribute to the conventional stability required by developments in the nuclear arms-control field. Conventional stability must ultimately come from structural adjustments to the military forces of NATO and the Warsaw Pact. Unfortunately, the negotiations specifically concerned with such adjustments, the MBFR talks, were in flux throughout 1986.

Over the years the MBFR negotiations in Vienna have been stymied by East–West differences over the issues of data exchange and verification measures. In December 1985 NATO attempted to finesse the deadlock with a new offer. This reversed the Alliance's long-standing requirement that agreement on data must precede reductions, proposing instead reductions in ground forces (5,000 for the US and 11,500 for the USSR) in exchange for a verification regime that would include a detailed data exchange and 30 inspections per year to verify troop levels and monitor compliance.

In February 1986 the Warsaw Pact replied by tabling a proposed draft treaty which assembled positions from three of its previous proposals, rather than responding directly to the NATO offer. In doing so it created significant doubt as to whether the Pact would ever agree to conventional force reductions at MBFR. These doubts were exacerbated by Gorbachev's April speech to the East German Communist Party congress, calling for substantial reductions in the land and tactical air forces of the European states, as well as Canadian and US forces in Europe (later explained as 100,000–150,000 within one or two years, and 500,000 by the early 1990s). More important, Gorbachev expanded the area to be covered to 'the entire European territory from the Atlantic to the Urals' and suggested that 'operational-tactical nuclear weapons could be reduced simultaneously with conventional weapons'. Hinting at some flexibility in the previously inflexible Warsaw Pact position, he stated that 'both national technical means and international forms of verification, including, if need be, on-site inspection, are possible'.

On 11 June, after a meeting in Budapest, Warsaw Pact leaders issued a statement filling in details of Gorbachev's April offer. They suggested that an appropriate forum to consider these proposals could either be the second phase of the CDE, a new negotiation (which would include all European states, Canada and the US), or an expanded MBFR framework. The Gorbachev initiative and the Pact statement did not in themselves constitute substantive progress in conventional arms control, but they did indicate that a new framework, which would broaden negotiations to include part of the Soviet Union, might be a catalyst for Pact flexibility. In his April speech Gorbachev had said that 'we believe that the formulation of the question in a broader context will make it possible to cut the knot which has been growing tighter at the Vienna talks over so many years now'.

The Warsaw Pact's failure to respond to NATO's December 1985 proposal, Gorbachev's April speech, and the 'Budapest call' have resulted in a NATO reappraisal of conventional arms-control talks. Meeting in Halifax, Nova Scotia, on 29–30 May, the Alliance's foreign ministers agreed to establish a High Level Task Force on conventional arms control to consider whether new approaches might better meet the options that had now appeared. These deliberations included France, which had deliberately taken no part in the long-running MBFR negotiations but would clearly have a major involvement in any structural arms control embracing the enlarged 'Atlantic-to-Urals' area. In the end, the task force did not live up to its original mandate to 'seek bold new steps', but NATO foreign ministers did issue the Brussels Declaration on Conventional Force Reductions in December 1986, recommending new talks on conventional arms reductions covering the whole of Europe and a separate negotiation that could build on the CDE approach.

Ultimately, the fate of conventional arms control 'from the Atlantic to the Urals' – and, by inference, the fate of MBFR – will be influenced by the 35-member CSCE currently meeting in Vienna. The NATO countries hope to establish a new negotiation on conventional arms control that will expand the focus of MBFR to all of Europe and will involve all the 23 Warsaw Pact and NATO states individually, rather than as bloc members. A precondition for such an expanded negotiation will be a mandate from CSCE for a second phase of the CDE to continue the work of the Stockholm meeting. It is not clear how the Warsaw Pact will respond to these conditions.

If new conventional arms-control negotiations based on these principles are convened, further modest and incremental improvements in the field of operational arms control – through increased openness about military activities, and possibly through constraints on deployments and activities – should be achievable. But, while it should be possible to tackle structural arms-control issues more logically in an enlarged 'Atlantic-to-Urals' forum, the issues which stultified the more narrowly focused MBFR negotiations (data agreement and verification measures) appear to remain stumbling blocks. More importantly, even if these could be overcome, it is difficult to see any realistic possibility that the West – lacking, as it does, any significant bargaining counters – could somehow negotiate away much of the Warsaw Pact's conventional advantage in Europe.

Chemical Weapons

The Conference on Disarmament continued to work towards agreement on a chemical weapons ban. At the start of 1986 there existed broad agreement on a general prohibition on the acquisition, production, possession and use of chemical weapons; the elimination of existing stocks and production facilities over a ten-year period; controls on remaining chemical facilities to ensure that no new chemical weapons are produced; and the establishment of a consultative committee to oversee any convention, including any verification activities. Disagreement still existed on a number of points, but the most difficult was that of verification measures – specifically challenge inspections. The US insists on short-notice (48-hour) mandatory challenge inspection procedures that would allow US or Soviet inspectors to check non-compliance anywhere on national territory. The USSR, although reconciled to the concept of on-site inspections, maintains that automatic challenge inspections could take place only at the declared chemical weapons production facilities, or in cases where the use of chemical weapons has been alleged. No significant flexibility in these positions has developed, although the US and the Soviet Union held bilateral discussions on a chemical weapons ban, as well as on a chemical weapons non-proliferation treaty, throughout the year.

Nuclear Test Ban

Although largely overshadowed by the other developments in bilateral negotiations, the issue of nuclear testing was an active part of the arms-control dialogue throughout 1986. The USSR continued to observe its unilateral moratorium on nuclear tests, begun in August 1985, as well as to call for a comprehensive test ban (CTB). While the Reagan Administration maintained that this was an ultimate goal of US policy, it remained opposed to a CTB on the familiar grounds that, so long as the West depended upon nuclear weapons for deterrence, it needed to continue to test them, both to ensure weapons reliability and to modernize its stockpile. Also, the Administration believes that a CTB cannot be satisfactorily verified. Consequently, the US goal in 1986 remained the implementation of the 1974 Threshold Test Ban Treaty (TTB) and the 1976 Peaceful Nuclear Explosions Treaty (PNE), but only after new negotiations had established verification measures going beyond those in the Treaties.

Notwithstanding the opposing US and Soviet positions on nuclear testing, a series of bilateral discussions were held in Geneva in July and September to explore verification improvements, which would be critical to monitoring any agreement. They were punctuated by Reagan's announcement on 10 October – in response to pressure from the Senate – that he would submit the TTB and PNE to the Senate for consent and ratification. The US–Soviet dialogue on nuclear weapons testing culminated in the Reykjavik meeting, where discussion of a phased reduction of tests and an eventual ban complemented the concurrent discussions on other issues. Soviet and US accounts of the meeting differ on how far the two sides' positions on nuclear testing converged, but any progress that was made fell by the wayside, like all else, with the final disagreement over SDI. A third bilateral discussion took place in Geneva in November, but no progress resulted, and the US and Soviet positions remained similar to what they had been at the beginning of the year. Prospects for improvement are marginal. The Reagan Administration still seeks to get the Senate to ratify the TTB and PNE, but with the reservation that the President would implement the treaties only when effective verification had been assured. In addition, Gorbachev's unilateral testing moratorium expired on 1 January 1987, and, after two US explosions early in the new year, a Soviet test took place on 26 February.

Outlook

At the end of a dramatic year in arms control, questions remain. The most basic is whether the Reagan Administration will work towards a compromise agreement on strategic defences, based on elaboration or redefinition of the ABM Treaty, in exchange for unprecedented reductions in strategic offensive weapons. Or will both sides be content to give priority to reaching an INF agreement and leave it to the next US Administration to make the strategic choices involving the future of offensive and defensive forces?

Another set of questions concerns the objectives that should be pursued in future conventional arms-control negotiations. How far should the West press in CDE for constraints on the military activities of both sides, and what kinds of measures should be agreed to reduce the secrecy of military activities? And can the West expect any significant progress in the follow-on negotiations on conventional force reductions, so long as the Warsaw Pact maintains a significant advantage in conventional military power?

The link between nuclear and conventional arms control was strengthened in 1986, but, distracted by procedural issues, the West has yet to decide on its objectives in the conventional arms-control negotiations. The new format for these negotiations might provide the basis for promoting conventional stability in Europe. However, success will depend on resolving difficult political issues, involving deadlock over data exchange and verification, as well as designing objectives which recognize the fact that – given the current balance of conventional forces in Europe – NATO has little bargaining leverage, if any.

The Super-powers

In 1986, as in the previous year, super-power relations were again determined by the two countries' leaders. The face-to-face confrontation in Reykjavik was a concrete symbol of the extent to which their personalities have come to dominate the super-power dialogue. By the end of the year it was clear that, over the short term, the course of super-power relations would be fundamentally affected by the struggles of Reagan and Gorbachev to maintain their authority in the face of serious domestic political difficulties.

Perhaps the most dramatic development was the precipitous decline in the popularity of President Reagan, and its profound implications for his ability to shape US foreign policy. Having begun the year in the afterglow of the Geneva summit, with his popularity at a record high, he ended it with his authority considerably dimmed. Despite successfully managing the public reaction to the failure to reach agreement in Iceland, the President found his political power diminished by substantial Democratic gains in the Senate elections in November, and his credibility was deeply tarnished by the Iran/Contra fiasco. His own Cabinet sought to distance itself from the discredited arms sales to Iran and the diversion of funds to the Contras, and a steady stream of key advisers began to leave his Administration. An emboldened Congress seemed ready to snatch control over security policy and to thwart the President's ability to pursue his chosen course on key issues of strategic arms and SDI. Unless Reagan can take bold steps to reassert his leadership (and an INF agreement may help him) he runs the risk of being reduced to a caretaker President.

Gorbachev's position also seemed uncertain as he began increasingly to seek international publicity and to pursue international initiatives to bolster policies that were meeting resistance at home. His speeches were sometimes censored, and initiatives on which he had staked considerable prestige were pruned back. Whatever the dangers from within, however, the public contest between the US and the USSR in 1986 clearly left Gorbachev in the stronger position. He was able to pursue an ambitious foreign-policy agenda that included a blizzard of arms-control proposals, a nuclear test moratorium that extended for well over a year, and the meeting with Reagan at Reykjavik. His Vladivostok speech in July heralded a new round of attempted openings to China, Japan and other East Asian states, and his successful visit to India demonstrated his remarkable sense of how to use the international stage. Gorbachev's proposals to end the war in Afghanistan, and the dramatic, unforeseen release of prominent dissidents (Sakharov and over a hundred others) may simply be public relations ploys, but they were well calculated to disarm Western critics by removing some of their most powerful anti-Soviet ammunition.

Despite Gorbachev's unquestioned drive, few of his foreign-policy initiatives have actually born fruit. He may have hoped to persuade Europeans that the failure in Reykjavik was due to American intransigence, but he was only half successful; in truth, many European governments breathed a sigh of relief that some of the proposals were *not* agreed. Symbolic troop reductions in Afghanistan and Mongolia were dismissed as a sham. His high foreign-policy visibility was a risky venture, for, if his efforts failed, his leadership of the Party could be steadily weakened.

Throughout the year the Soviet leader seemed to be at pains to resist opportunities to heighten tensions with the United States. Despite pointed US rejection of the Soviet nuclear testing moratorium, the USSR persisted with it for over a year, and when Moscow announced its intention to resume nuclear testing after the first US test of 1987, it was made clear that a resumption of the moratorium remained possible if the US would reciprocate. The Soviet response to the US bombing of Libya was highly muted. Even the Zakharov/Daniloff incident, which threatened to escalate into a serious confrontation, was defused by the intense round of negotiations between Foreign Minister Shevardnadze and Secretary of State Shultz, leading to the release of Zakharov, Daniloff and the prominent dissident Yuri Orlov and the agreement to hold the Reykjavik meeting. Gorbachev seems determined to keep doors open, if only to avoid public-relations disasters similar to the Soviet walk-out from the arms talks in late 1983.

The summit itself seemed to epitomize the current ambivalence in the super-power relationship. To have obtained President Reagan's agreement to the meeting must be seen as masterful tactics by Gorbachev, who not only managed to find a way out of the Daniloff problem, but skilfully dodged his commitment to follow the Geneva meeting with a summit in the US that seemed unlikely to produce the kind of agreement that would justify such a gamble with his prestige. Although the US sought to play down the significance of the meeting (the US called it a 'base camp' to a summit) Gorbachev's opening gambit of a comprehensive arms-control proposal put Reagan on the defensive. Despite two days of intense discussions, neither side was willing to make the key compromises needed for a breakthrough, and the two departed frustrated, even angry. Let slip in the confusion was what appeared to be a very real opportunity to conclude significant agreements on intermediate-range nuclear forces (INF) and perhaps on crisis management procedures.

Failing to achieve any agreement at the summit, both sides sought to dispel the gloom by stressing the progress made and their hopes for the future. President Reagan said: 'we have it within our grasp to move speedily with the Soviets toward even more breakthroughs'. And Gorbachev, too, seemed prepared to press forward, rather than wait for a successor in the White House. The appointment of Soviet First Deputy Minister Yuli Vorontsov to head the Geneva negotiations reinforced this message. On balance, it seemed that Gorbachev, pursuing support for his domestic reform efforts, would benefit from an agreement with the United States if it demonstrated that he had won significant concessions. Dealing effectively with the American President would cement his own legitimacy, much as the SALT I agreements had bought Brezhnev the political capital to pursue detente in the 1970s.

Whether any significant progress in US–Soviet relations can be achieved in the near future will depend critically on the course of events in Washington during the coming months. It remains to be seen whether President Reagan will perceive an arms-control agreement as a way to rescue his teetering Presidency. This in turn will depend on whether he is in fact prepared to be flexible on SDI, despite intense pressure from conservatives in the US to accelerate the pace of his efforts to deploy strategic defences. To reach an arms-control agreement the President will have to unite, or at least override, the warring factions in his Administration and restore some coherence to the decision-making process which has come unravelled in the wake of the Iran/Contra affair.

Without some progress on arms control a further meeting between Gorbachev and Reagan is unlikely, and US–Soviet relations may well move from quiescence to quarrelling. Both countries seem likely to be preoccupied with internal politics, as the US prepares to enter the long process of choosing a successor to Reagan, and Gorbachev continues his effort to gain a firmer grip on the reins of power. The seeds of difficulties lie in this preoccupation, however, for foreign policy will be an important part of the electoral debate in the United States, while in the Soviet Union a firm stance *vis-à-vis* the US may be required to ensure the support of the conservatives and the military.

THE UNITED STATES: PRESIDENCY IN CRISIS

In the history of the Reagan presidency, 4 November 1986 was a fateful day. In the United States, voters in the Congressional elections decisively rejected the President's efforts to maintain Republican control of the Senate, turning out of office seven incumbent Republicans, many of them the President's most loyal supporters, and handing the Democrats a ten-seat majority. Meanwhile, thousands of miles away, the Speaker of the Iranian Parliament, speaking on the seventh anniversary of the seizure of the US Embassy in Tehran, recounted a tale of secret US arms sales to Iran designed to secure the release of American hostages in Lebanon. The bizarre story – which began almost comically, with accounts of autographed bibles, forged Irish passports and a cake in the shape of a key – quickly turned into the most profound crisis of Reagan's presidency, posing a challenge to his ability to lead the nation through the last two years of his second term.

The revelation of these arms sales, which first appeared in a small, pro-Syrian Beirut weekly, *Al Shiraa*, immediately embroiled the Administration in an intense controversy. The American hard line on not negotiating with terrorists, which Secretary of State George Shultz had pressed with such fervour on the European allies, came to be seen by many as a duplicitous cloak for the United States' own dealings, and the Administration's credibility suffered severely both at home and abroad. But there was worse to come. Three weeks later, the President and his Attorney General, Ed Meese, appeared before a hastily-summoned Washington press conference to announce their discovery that some of the proceeds of the arms sales (estimates ranged from $4 million to perhaps as much as $30 million) had been diverted to the Nicaraguan Contras, at a time when Congress had barred military aid to them. In the space of a few weeks the President's National Security Adviser, Admiral John Poindexter, and Lt Col. Oliver North, an architect of the Iran/Contra fiasco had both resigned; a new National Security Adviser, Frank Carlucci had been appointed; the ailing CIA Director William Casey (who had been implicated in the arrangements with Iran) had been replaced; and a special prosecutor was named to investigate possible criminal violations. Although his days were numbered, White House Chief-of-Staff Donald Regan continued to cling tenaciously to office, despite calls by many, including Congressional Republicans, for his resignation in the light of suggestions that he had authorized the Contra aid.

Details, often contradictory, continued to emerge about the nature of the covert transactions as Congress created two investigatory panels to look into what came to be known variously as Irangate, Iranamok or Iranagua. The arms transactions seem to have begun with Israel selling to Iran equipment of US origin from its own inventory in August 1985; this was shortly followed by the release of the Rev. Benjamin Weir from captivity in Beirut. The first phase of the dealings apparently ended in December 1985. Then, on 17 January 1986, over the objections of Secretaries Shultz and Weinberger, President Reagan issued a secret finding authorizing the sale of US equipment to Iran by the US itself, using Israel as an intermediary; it was specifically directed that Congress should not be informed of the Administration's intentions. In May former National Security Adviser Robert McFarlane made his now famous trip to Tehran with the bible, the cake and a planeload of spare parts. According to McFarlane, he expected the delivery to be followed by the release of four hostages, but none was freed. A further shipment in July was followed by the release of the Rev. Lawrence Jenco, and after additional shipments of 500 *TOW* missiles on 31 October, David

Jacobson, another hostage, was set free on 2 November, just two days before the US elections.

Despite the apparent linkage between the arms deal and efforts to secure the hostages' release, the President insisted that the contacts with Iran, including the arms transfers, were designed to strengthen the position of so-called Iranian 'moderates' in the domestic struggle for the succession to Ayatollah Khomenei. But even the Administration's own experts were sceptical of the wisdom of such an attempt to influence Iranian internal affairs, and a majority of Americans doubted President Reagan's veracity. The President strove to put the issue behind him (acknowledging in his State of the Union address that 'serious mistakes were made', but insisting that he 'did not believe it was wrong to try to establish contacts with a country of strategic importance to try to secure freedom for our citizens'), but the controversy would simply not die away. In early January 1987 the new Congress voted to set up formal investigations. On 26 February the Administration suffered another blow with the release of a report on the affair by a panel of 'three wise men' (former US Senator John Tower, former Secretary of State Ed Muskie and former National Security Adviser Lt Gen. Brent Scowcroft), appointed by the President in late November.

The Tower Commission concluded that, 'whatever the intent, almost from the beginning the initiative became in fact a series of arms-for-hostages deals', and, although it found no evidence of wrongdoing by the President, it criticized him for his failure to 'force his policy to undergo the most critical review'. It focused most of the blame on Poindexter (who 'failed grievously on the matter of Contra diversion') and North, but also singled out Donald Regan ('he must bear primary responsibility for the chaos that descended on the White House'). It also criticized Shultz and Weinberger for distancing themselves from the unfolding events and failing to protect the President from the consequences of his commitment to free the hostages.

Regan's resignation followed almost immediately upon the publication of the report, and he was replaced by former Republican Senator Howard Baker. Baker, a widely respected former Senate Majority Leader, had gained a national reputation during the Watergate hearings in 1973. Many hoped that he would be able to provide the leadership necessary to extricate the White House from the chaos that had followed the Iran/Contra disclosures.

The Iran/Contra scandal left a good deal of wreckage in its wake, over and above the resignations and firings that it prompted. US anti-terrorism policy had been discredited, handicapping US efforts to respond to a new wave of kidnappings in Lebanon in early 1987. US–Israeli relations were sorely strained, as each side tried to put the onus for initiating the arms transfers on the other. The prospect for continued military aid for the Contras – already in doubt as a result of the new Democrat majority in the Senate – suffered a severe setback, even though there were doubts about how much of the allegedly diverted funds actually reached the intended recipients. Although Secretary Shultz had managed to distance himself somewhat from the debacle, uncertainty remained over his conduct of American foreign policy and his continuing effectiveness. Perhaps most important, the President's competence and leadership came under intense scrutiny, and many questioned whether age and illness might not have weakened his ability to pull the Administration out of the crisis. It had already been a difficult year for the President in his continuing struggles with Congress over the budget, arms control, trade policy and South Africa. As the criminal and Congressional investigations continue well into 1987, he will be hard pressed to seize the initiative from a Congress that seems increasingly inclined to go its own way.

Budget Battles

The defence budget story in 1986 was essentially a repeat of the previous year's saga. On 5 February the President submitted his budget to Congress, including a request for an 8.2% real increase in overall defence spending (from $286.1 billion in FY 1986 to $320.3 billion in FY 1987). Under the President's plan, defence would grow from 27.1% of the federal budget to 28.4%. But, to an even greater extent than before, the President's proposals were simply ignored by a Congress which, for the first time, was operating under the demanding strictures of the Gramm/Rudman/Hollings budget balancing legislation, enacted the previous year. Under this legislation, Congress faced the prospect of having to reduce the overall deficit from $212 billion in FY 1986 to $144 billion for FY 1987. It was apparent from the outset that defence spending, a major component of the 'controllable' federal budget (those items which are subject to annual appropriation – unlike, for example, interest on the debt or Social Security), would take an important share of the required reductions. Although the Gramm/Rudman enforcement mechanism was ruled unconstitutional by the Supreme Court in July, the political pressure for deficit reduction led Congress to try to meet the targets which that legislation imposed.

Despite the general consensus on significant cuts in the President's proposals, the budget process in Congress nearly broke down entirely in 1986. Both Houses passed their original budgets in the Spring, but disputes over how to achieve the required deficit reductions held up final approval of a compromise budget until late June – more than two months after the time formally mandated in the Budget Act. For the first time, Congress was unable to pass a single appropriation bill before the end of the fiscal year, leading to yet another serious confrontation between President and Congress, and even a brief shutdown of the government when the President refused to accept any further short-term extensions of US spending authority.

For the defence bill itself, a major source of difficulty was the growing antagonism between Congress (particularly the House of Representatives) and the President over arms control. A number of things heightened the conflict during the year – most notably the Administration's continued refusal to respond positively to the Soviet moratorium on nuclear testing, and the May announcement that the Administration would no longer continue its policy of 'not undercutting' the SALT II Treaty as new systems (*Trident* submarines and air-launched cruise missiles on B-52 bombers) came into inventory. The version of the defence bill adopted by the House included a number of provisions challenging the Administration's policy: including measures that would require the President to adhere to the SALT II Treaty; prohibit testing of its F-15-launched anti-satellite weapon against an object in space; forbid US nuclear tests of above 1 kiloton so long as the USSR continued its moratorium on testing; and ban further development of binary chemical munitions.

With the House more determined than ever to achieve at least some of its arms-control agenda, a stalemate seemed possible. But the pressures of the impending Reykjavik summit (and the President's not-so-subtle hints that he was prepared to blame Congress for limiting his freedom if the summit failed), plus Congress members' eagerness to campaign for the November election, broke the log-jam. The House gave way on most of its provisions (except for the ban on ASAT testing). When the dust finally settled Congress had approved a defence budget of $292 billion ($28 billion below the President's request). But actual defence appropriations (included in an unprecedented omnibus continuing appropriations resolution containing $576 billion to fund virtually the entire US government) reached only $289 billion. Although the Administration's FY 1988 budget proposal for 3% real growth ($312 billion) is the smallest of the Reagan presidency, it is still unlikely to satisfy Congress, and the prospects seem poor for any real growth in defence spending for the foreseeable future. Congress has already made clear that it is unlikely to approve the Administration's $2.8-billion supplemental defence spending request for FY 1987, which was designed to make up the difference between the level set in the budget passed by Congress and the actual appropriation.

Despite the significant cut in overall spending, Congress followed its familiar pattern of reducing purchases and prolonging programmes, rather than cancelling major systems. But with the prospect of little or no growth in defence spending for several years ahead, both the Pentagon and Congress will increasingly be forced to make more dramatic choices. The FY 1988 budget, presented in January 1987, shows some evidence that the cuts are beginning to bite. The Air Force, for example, has formally abandoned its goal of 40 air wings (it plans only 37), and the Army has been forced to terminate several helicopter production lines prematurely.

Foreign Aid

The Administration's budget problems were not confined to the Defense Department. The President requested a $1-billion increase in foreign aid (including both economic and military assistance), from $14.4 billion in FY 1986 to $15.5 for FY 1987. Despite urgent pleas from Secretary of State Shultz who warned that cuts in the proposal would be 'a tragedy for US foreign policy', Congress turned a deaf ear, slashing the funding to $13.4 billion. Because key countries (particularly Israel and Egypt) were held immune from the cuts, and others (Greece, Turkey, Cyprus, Pakistan and Northern Ireland) were protected from the most severe cuts, many of the remaining countries will face reductions of the order of 30%–50%. The reductions threaten to pose especially serious problems with countries where the US is involved in negotiations over military base rights (such as Spain, Portugal, Greece and Turkey). Congress did, however, respond to one compelling case – voting an additional $200 million (beyond the amount included in the general appropriation bill) in aid to the Philippines.

SDI

Wrapped up in the budget wrangle was the intensifying struggle between Congress and the President to determine the scope of the Strategic Defense Initiative (SDI) programme, and the overall goals of the US in the area of ballistic missile defence. The Administration's original FY 1987 budget proposed $5.3 billion for SDI (including Department of Energy funds for SDI-related research), a 73% increase in funding over the previous year's total of $2.7 billion. Although Secretary Weinberger, in his annual report to Congress, maintained that SDI 'is not a weapons development program, nor is it a program with preconceived notions of what a potential defensive system against ballistic missiles should entail', it was clear that the Administration, through its budget requests and attempts to narrow the range of likely technologies, was seeking to move SDI into a higher gear. Several highly visible demonstrations (including an experiment in September in which one satellite successfully tracked and collided with another target satellite) were trumpeted as signs of the growing maturity of the programme. And the programme managers seemed to be getting closer to deciding on the types of technologies (such as ground-based free-electron lasers, neutral-particle beams for mid-course tracking and discrimination, and kinetic-kill homing rockets for boost-phase intercept) which seemed most promising for the actual deployment of systems in the 1990s.

But Congress seemed increasingly discontented with the pace and direction of the programme. The Senate, most supportive in the past, cut the President's proposal to $3.9 billion and only narrowly (by a single vote) defeated an effort to hold SDI spending down to 3% real growth. Perhaps even more important, the Senate Report on the bill included a direct challenge to the stated goal of a population defence, declaring that 'the major emphasis within SDI should be dedicated to developing survivable and cost effective defensive options for enhancing the survivability of US retaliatory forces and command, control and communication systems'. The final compromise between the Senate and the House (which had originally approved only $3.1 billion) provided $3.5 billion for FY 1987 (a 28% real increase over FY 1986), and the Conference Committee pointedly endorsed the Senate's advisory language over SDI goals.

In the wake of the Reykjavik summit, the Administration mounted a massive effort to convince the American public that the President was right to stand firm in defence of the SDI programme, and the campaign appeared to be effective, with polls showing that a substantial majority of Americans approved the President's stand. However, an attempt to mobilize this opinion in favour of Republican senatorial candidates failed, and a number of incumbent Republican senators, who had formed an important element of Reagan's support in Congress, lost to Democratic challengers.

Thus the FY 1987 compromise over SDI simply presages even more controversy to come. By the beginning of 1987, supporters of SDI, including Secretary Weinberger and prominent conservatives in Congress, began a major effort to insulate the programme against the possibility that Reagan's successor might be less sympathetic to its goals by committing the US to an early deployment of rudimentary ballistic missile defences (BMD) and abandoning the commitment to keep the programme within the 'narrow' interpretation of the ABM Treaty. This effort is certain to cause deep controversy within the Administration, with ABM Treaty supporters in Congress, and with America's allies, who see it as undermining any remaining hopes of reaching a strategic arms reduction agreement with the USSR.

Closely related to SDI is the US space programme. The explosion of the space shuttle *Challenger*, followed by the failure of *Titan* 34-D and *Delta* launchers, crippled the US ability to place vital military payloads, including surveillance satellites, in space. The delays in restarting the Shuttle programme, and the fierce competition for precious places in the payload queue over the next few years, could raise problems for the pace and timing of SDI experiments, not to mention any attempt to deploy space-based BMD assets in the near future. A sign of the concern was Secretary Weinberger's decision to request $110 million, as part of the Administration's supplemental appropriation request for FY 1987, for research and development for a heavy-lift launch vehicle for SDI.

The Nuclear Forces

1986 was a milestone year for the US strategic forces. After more than a decade of debate two new strategic systems – the MX multiple-warhead ICBM and the B-1 bomber – were declared operational. But, given the past histories of these systems, it seemed only fitting that the deployments should be accompanied by a new round of controversy, rather than seeing the issues laid to rest.

The first months of the B-1 could hardly be called auspicious. Even during flight testing, it had been plagued by continuing problems of fuel leakage (the Pentagon insisted it was not leakage, only 'weeping'). Then, as the deployments began at Dyess Air Force Base in Texas, a whole range of problems began to emerge, including difficulties with the terrain-following radar, electronic countermeasures equipment, flight controls and missile launching systems. Many critics contended that an important source of difficulty was the 41-ton weight increase between the bomber's initial design as the B-1A and its actual deployment as the B-1B, and that the resulting higher wing loading and lower thrust-to-weight ratio were adversely affecting its low-level performance and range. The Air Force asserted that many of the problems were to be expected in the early years of operating a complex new aircraft, but its request for an extra $600 million for flight testing and work on the electronic systems suggested that they were not insignificant.

The MX, by contrast, had a smoother technical (if not political) path to deployment. But the Administration seemed determined to keep alive the prospect of further deployments beyond the 50 approved by Congress; at the end of 1986 it proposed that 50 more should be deployed on railroad cars, to be kept at military installations in peacetime but deployed on the national rail system at times of crisis or 'strategic' warning (the so-called 'rail garrison' basing option). Preliminary estimates put the cost of developing the basing system at $2.5 billion, plus $4–5 billion more to procure the cars and associated equipment. This proposal was promptly opposed, both by those who doubted its wisdom or viability and by those who saw it as yet another effort to undermine the mobile, single-warhead *Midgetman* ICBM.

Midgetman itself was the subject of intense debate during 1986, fuelled by doubts as to whether the Administration truly supported the project (notwithstanding the inclusion of $1.2 billion in the Defense Department budget to begin full-scale development of the missile). These concerns stemmed from the Administration's arms-control proposals (which sought a total ban on mobile missiles unless the USSR could demonstrate that numerical limits could be verified) and from suggestions by senior Administration officials that *Midgetman* might be fitted with two or three multiple, independently-targetable re-entry vehicles (MIRV). For the moment, however, the Administration has approved the single-warhead version. Behind all the substantive disagreement lurked a doubt as to whether funds would in fact be available to foot the expected $50-billion bill for the 500 *Midgetmen* planned.

Military Reorganization

On at least one front Congress and the President did reach an important agreement in 1986: reform of the US military organization. The Administration had previously resisted the efforts of military reform advocates in and out of government, but the March 1986 report of the Packard Commission, which called for significant changes, seemed to tip the balance, and soon afterwards the President announced his support for at least some of the measures that had been on Congress' agenda for a number of years. With Presidential backing assured, Congress went to work and quickly produced a package which was passed unanimously in both Houses. The legislation substantially strengthened the role of the Chairman of the Joint Chiefs of Staff, created a new position of Vice Chairman of the Joint Chiefs, and increased the authority of the unified commanders, providing them with a stronger role in the budgetary process and in controlling the activities of the individual service elements under their command. The problem is that such procedural changes only partially relax the individual services' grip over the forces that will be made available to operational commanders and over budgets, which raises questions about how effective reform in fact will be.

Military reformers also turned their attention to the continuing service rivalries which have beset US special operations forces. The planning and operational confusion that had plagued the Iran rescue mission and the use of special forces in Grenada led Congress to create two new positions – an Assistant Secretary of Defense for Special Operations and Low Intensity Conflict, and a Commander of Special Operations to whom all special forces stationed in the US would be assigned. The latter would have authority for training, strategy and tactics, and responsibility for participating in the development of budget proposals for the special operations forces in all services; operational command would normally remain with the unified Commander in the geographic region in which the forces were used, but the new special forces Commander could be designated to carry out operations directly. This Congressional foray into dictating military organization was virtually unprecedented, but the Members concluded that 'the seriousness of the problems and the inability or the unwillingness of the Department of Defense to solve them left no alternative'. The Pentagon has gone ahead with creating the new command, but awaits further Congressional authorization of the new Assistant Secretary post.

Regional Issues

The President's problems with Congress were not limited to military issues. He did manage to score one notable success, when Congress reversed its earlier ban on military assistance to the Contras; a dramatic vote in the House of Representatives on 25 June approved, by 221 to 209, $70 million in military assistance and $30 million in humanitarian aid, after the President's request had been voted down by a similar margin in March. But this victory is likely to prove short-lived. The President reiterated his support for the Contras in the strongest terms in his State of the Union address in January 1987, but the revelations of the diversion of funds from the Iran arms sales, another round of internal bickering among the Contras, and the Democrats' acquisition of control of the Senate, all make the task of maintaining financial support for the Contras a daunting one indeed.

The Administration also continued to press for support for other prongs of the 'Reagan Doctrine', including the sale of *Stinger* air-to-surface missiles to rebel forces in Angola and Afghanistan. The problem was that the US could help to keep the rebels fighting but did not have the means to ensure that they would win. The *Stinger* issue has also proved particularly controversial in Congress, where many fear the consequences of the missiles falling into terrorist hands.

The Administration suffered a serious setback in its attempt to hold the line on sanctions against South Africa, when the Senate joined the House in easily overriding the Presidential veto of a bill banning new investments and bank loans, barring a range of South African imports, ending South African airlines' landing rights in the US, and blocking exports of oil, petroleum products, munitions and nuclear technology. This contrasted sharply with the Administration's success in 1985, when the President's decision to impose a more modest package of sanctions by Executive Order had headed off a Congressional attempt to legislate tougher measures. The fight over sanctions opened up an important rift between the Administration and several key Republican legislators, because Senator Richard Lugar, then Chairman of the Senate Foreign Relations Committee, defied White House pressure and led the effort to overturn the President's veto.

Trade and Economic Policy

Trade relations between the US and its allies assumed a new prominence in the domestic political debate in 1986. Throughout the year Congressional leaders warned the President that mounting trade deficits (reaching a record $170 billion in 1986 and still climbing in the early months of 1987) and dislocations in the agricultural and key manufacturing sectors were generating increasing pressure for protectionist measures against US trading partners. In May a strong bipartisan majority in the House passed a tough trade bill, but the Republican leadership in the Senate successfully blocked further consideration of the legislation by the Congress. A bill imposing textile quotas was passed by both Houses but was vetoed by the President, and the veto was narrowly sustained by the House.

Hopes that the rapid decline of the dollar against the mark and the yen would begin to reverse the deficit faded, and trade and agriculture issues figured prominently in the November elections – contributing, for example, to the defeat of two Republican senators in the farming states of North and South Dakota. Senior Administration economic policy officials mounted a heavy campaign to persuade West Germany and Japan to take steps to stimulate domestic demand as a way to redress trade imbalances, but this effort has so far achieved only modest success (in the form of a slight lowering of German and Japanese interest rates).

The US ended the year with a shot across the bows of the European Economic Community, announcing its intention to impose substantial duties on EEC imports in retaliation for the loss of US grain sales resulting from Spanish and Portuguese entry into the Community. Draconian measures were temporarily averted in negotiations between the US and the EEC, but there remains a strong prospect that the Democrat-controlled Congress will press the Administration for further action. The new Speaker of the House of Representatives, Jim Wright, announced that trade policy would be 'the first imperative' of the 1987 Congressional session. These economic tensions are already showing signs of broadening into a more general deterioration of alliance relations, particularly with Canada, Europe and Japan.

Rebuilding the Administration

The trauma of the Iran/Contra revelations produced a dramatic upheaval in the Administration's security policy apparatus – which was already beset by intense internal struggles. The resignations of White House Chief-of-Staff Donald Regan, National Security Adviser Admiral John Poindexter and CIA chief William Casey removed three of the principal figures involved in the fiasco. Reports of disarray and conflict among the President's key advisers reinforced the impression that the Administration was sinking into ineffectiveness for the President's final years.

The appointment as National Security Adviser of Frank Carlucci (who had served in the CIA and the State and Defense Departments under both Democrat and Republican Presidents), and that of Howard Baker to replace Regan, provided some reassurance that the Administration might be able to restore order to the decision-making machinery. But the continuing Senate and House investigations seemed sure to keep the Administration on the defensive, and the dramatic decline in public confidence in the President's leadership makes the task of reasserting that leadership all the more problematical. Democrat control of the Senate will bring the well-respected Administration critic Senator Nunn, new Chairman of the Armed Services Committee, into the spotlight and substantially reduce the Administration's ability to influence Congressional consideration of potentially explosive issues such as trade, arms sales (a new package of sales to Saudi Arabia has already been announced), arms control and SDI. It will be difficult indeed for the President and his senior advisers to prevent the focus of the security debate in the US from shifting to the post-Reagan era, as the country begins to turn in earnest to the issues and personalities that will emerge in the run-up to the 1988 Presidential election.

THE SOVIET UNION: HINTS OF CHANGE

The key factor governing developments in the Soviet Union during 1986 was a protracted and sometimes bitter struggle over reform – the need for it, and the extent and pace of it. Mikhail Gorbachev, in his first year as Party General Secretary, had been able to score early successes, but he met increasingly stiff resistance once it became clear that his drive for efficiency and reform – indeed for a 'revolution from above' – would threaten entrenched privileges and interests.

Undoubtedly there was change. Some of it was small and apparently trivial: the admission that the USSR has drug and AIDS problems, the possibility of listening to heavy-metal rock music, and the existence of more fruit and vegetables on the Moscow markets. Yet there were also surprising gestures and initiatives: Andrei Sakharov was given permission to return to Moscow, many other political prisoners were released, the head of the KGB publicly condemned one of his units for illegal acts, and there was a flurry of foreign-policy moves. The predominant factor, however, remained the evidence that Gorbachev's drive to consolidate his authority had run into opposition and would require considerably more time before it could be considered a success. The battle over reform has clearly entered a crucial stage. Its outcome and eventual repercussions are still in doubt.

Pressures and Responses

The gargantuan task of steering the Soviet Union onto a new and more promising course was virtually imposed on the new leadership. Pressure, both internal and external, had been building up for too long for it not to be relieved. Ever since the last years of Brezhnev's rule Soviet economic performance had been sluggish at best, and uncontrollable outside factors had further worsened the picture: collapsing oil and raw-material prices, a substantial decline of the dollar (which remains the base on which most Soviet hard-currency earnings are determined), a drying up of financial markets, and the Reagan Administration's stiff opposition to exports of high technology to the USSR. In addition, a succession of bad harvests – averaging just over 180 million tonnes, compared with a target of 238 million tonnes (and reaching a nadir of 158.2 tonnes in 1981) – required immediate remedial measures.

At the same time demands grew, and with them the strains imposed on the Soviet system. The USSR was confronted with the formidable strategic and technological challenge of the SDI. Its attempts to prevent the deployment of US intermediate-range nuclear missiles and cruise missiles in Europe failed, and transatlantic strategic coupling was strengthened. And, although the security-policy consensus within NATO was shaken – by the INF debate, the US positions on SDI and the ABM Treaty, and the US bombing of Libya – the Alliance maintained surprising cohesion, as the Spanish referendum vote for continued membership and the re-election of the Bonn coalition underlined. A more assertive US (accused by the Soviet Union of 'neo-globalism') transformed Soviet acquisitions in the Third World from potential springboards into beleaguered garrisons. Japan gradually moved closer to a security dialogue not only with the United States but with NATO as well, and the USSR could no longer either doubt or ignore China's determination to modernize. Most worrying of all, at the very moment when the Soviet Union was struggling with its economic perplexities the West appeared to be on the verge of undergoing a new, computer-driven, technological revolution.

In short, Gorbachev had inherited the leadership of a USSR that was facing a triple challenge. First, the generation of domestic resources was too slow, and the potential for growth based on outside factors too limited and precarious, for the economy to be able to cope simultaneously with the four major tasks demanded of it: to generate the investment desperately needed for modernization and future growth; to meet the present demands of the armed forces, and perhaps those of a new major arms competition with the US; to afford the rising costs of the Soviet empire; and to guarantee a slow, but steady, increase in the Soviet population's standard of living. Secondly, in the military realm, there was an increasing risk that the massive investments in strategic offensive weapons and the upgrading of Soviet conventional forces would be devalued by Western moves and technological change. Thirdly, the USSR has begun to recognize that it is no longer enough to rely on military power to increase its political influence; it will need to be economically and technologically competitive if it is to maintain super-power status. In sum, the Soviet Union needed what Lenin called a *peredyshka* (respite) from competitive pressures so as to concentrate energy and resources on the formidable task of rebuilding the bases of its power.

It may be Gorbachev's most remarkable achievement as a Soviet leader to have grasped, and determined to deal with, these realities. He appears to have recognized that even if traditional answers are still available, they will no longer be adequate to the complex challenges his country faces. He has therefore advocated a potentially coherent set of answers and initiatives to respond to the problem.

At home, he has called for an extensive 'restructuring' (*perestroika*) of the ailing economy, as a basic precondition for preserving the Soviet Union's power and ensuring its future. To be effective, this must be accompanied by two supporting moves: first, a massive purge of the Soviet government and Party apparatus, not just to broaden the

the General Secretary's personal power base, but also, and more fundamentally, to lick the petrified bureaucracy into more professional and efficient shape; and, second, a policy of greater 'openness' (*glasnost*), so as to permit controlled debate and guarantee the minimum flow of information that is indispensable in order to overcome the legacy of the Brezhnev era. In pursuit of these objectives Gorbachev has made some astonishing moves for a Soviet leader: rehabilitating previously banned authors (Nabokov and Pasternak, among others), reducing the jamming of foreign radio broadcasts, allowing the showing of a film about abuses of the Stalin era, sharply criticizing Brezhnev's last years, and inviting Soviet artists who had defected to the West to return to perform in the USSR.

All the same, Gorbachev does not seem to be interested in altering the basic Soviet system. He is, however, determined to make it work again. He believes that if the economy is to find its rhythm again, then at the very least dead wood must be cut out, discipline encouraged, and the Soviet citizen provided with a minimum of legal security (witness the critique of the judiciary, the encouragement of criticism from the bottom, the purges and disciplining of members of the police and security apparatus, and the release of a large number of political prisoners). However, as the forceful suppression of demonstrations in Moscow by KGB thugs on 10–11 February 1987 showed, even such limited domestic reform will require time.

To gain this essential time, outside pressure must be reduced, at least temporarily. Gorbachev has therefore vigorously advocated 'new political thinking' in foreign policy. The USSR can no longer rely on military power alone to reach its objectives, so the new thinking not only recognizes that military, political and economic factors are intimately linked, but states that 'ensuring security is increasingly becoming a political task which can only be solved by political means'. Gorbachev has not decided to cut military expenditures *per se*, but he hopes to keep them under tighter control. As a result, the armed forces may have to accept real short-term reductions in order to encourage economic growth – yet they should also be the main beneficiaries of growth in the longer term.

Nor has Gorbachev decided to reduce the military burden by seeking an arms-control agreement with the US at any price. But Soviet military power must be married to political and arms-control initiatives, and this requires a new style and purpose in Soviet diplomacy. The ultimate objective of this approach is still not clear. Gorbachev is pursuing arms control with the West, but none of his proposals represent drastic changes from past objectives, and they still protect the major elements of Soviet military power. It remains interested in delaying or killing weapons programmes perceived as particularly threatening to Soviet interests through political means (the 'peace struggle' in a united front with Western 'peace-loving forces') and in weakening the cohesion of the Western Alliance.

In essence, Gorbachev appears determined to forge Soviet economic, defence, foreign and arms-control policies into a coherent national security policy, both to overcome the present crisis and to lay the groundwork for future growth of Soviet power and influence. He is without doubt the most formidable Soviet leader the West has had to cope with in a long time. But first he must implement his domestic agenda, and this has proved a daunting task.

Domestic Reforms and Resistance

Gorbachev's domestic policy initiatives in 1986 focused on five interlinked key areas: the streamlining of the economy, Party reorganization and cadre policy, the purge of provincial fiefdoms, the campaign against corruption and sluggish work morale, and a carefully controlled move towards a limited liberalization of Soviet society. The rhetoric escalated as the year moved on, but actual progress was slow, and resistance was growing.

Nowhere was this so apparent as in the drive to reform the economy. In a major speech in Khabarovsk on 31 July, Gorbachev declared that there were no ready formulas for speeding up economic growth. What was necessary was a reorganization of a long-term and open-ended nature – not simply a reform, but a 'revolution'. This message could be read as implying that the system might not only be improved but fundamentally altered – keeping open even the option of reverting to a market economy in certain sectors as a last resort. Little action was seen that would substantiate such a view, however. Steps were taken to foster agricultural output by somewhat liberalizing the regime under which agricultural co-operatives could sell produce on the markets; and in November a 'law on individual labour activity' was promulgated, legalizing, but also regulating and taxing, some of the widespread moonlighting in the service and crafts sector. But these modest moves were counterbalanced by the adoption of harsh and highly unpopular regulations against so-called 'unearned incomes' (on which many Soviet citizens actually depend).

The main message of Gorbachev's reform policy was not a questioning of the existing centralized system. If there was an increase in the autonomy and decision-making powers of economic entities, this was accompanied by stiffer quality controls and a sharpening of the central control and decision-making apparatus. Clearly, central planning is expected to remain the guiding principle and is not to be replaced by the law of supply and demand. Efficiency is to be increased by such means as the creation of 'super-ministries' dealing with key sectors of the economy – agriculture, machine-building, energy and foreign trade.

The day-to-day working of factories, farms and research institutes continued to be affected more by disciplinary pressures than by organizational changes that might increase motivation. What change there was tended to be superficial and administrative; it was rarely bold. Even so, change was seen by many as a threat to the existing patterns of their lives: wages linked to productivity may hold out promise for some, but for many it will mean that they earn less, unless they work much harder and more efficiently. Truly far-reaching reform, let alone a 'revolution', would inevitably call into question the social hierarchy and stability upon which Soviet society rests. The clinging to entrenched privileges – most pronounced in the *nomenklatura*, yet ultimately reaching far beyond it, since almost everybody has his niche of corruption on which he depends – turned out to be a formidable obstacle to Gorbachev's reform drive. This provides the background to the General Secretary's ever more desperate complaints about the hidden resistance which his policies were meeting during the year.

Attempts were made to bypass or to defuse the problem. Efforts to generate growth in income from abroad were made, and steps were taken to encourage a controlled debate (including criticism of inefficient *apparatchiks*) and to ensure a minimum of personal security indispensable for such a debate. The foreign trade monopoly was relaxed by permitting more than 20 ministries and 70 large production entities to engage directly in foreign trade activities as from 1 January 1987. Joint ventures with Western companies (which were allowed a 49% share) were reluctantly permitted in order to attract not only Western capital and technology but also marketing and managerial experience. Gorbachev repeatedly called on the population to bypass established channels and report wrongs committed by the governing apparatus directly to the top. But a change in atmosphere is necessary before any Soviet citizen will believe that he could indulge in such behaviour. The censuring of the judiciary and the security apparatus, as well as the loosening of the tight constraints in cultural affairs which Gorbachev has initiated, may contribute to the necessary change in environment and encourage some action.

Eventually, however, Gorbachev has to break the resistance if he is to ensure the continuation of his crucial drive towards reform. A purge of the inefficient, and of the relicts of the Brezhnev era, in both the top leadership and in the middle and lower levels of the bureaucracy was, and remains, his ultimate test. Yet it was here that he failed most strikingly during the year. Neither the Central Committee (CC) plenum in June, nor the two sessions of the Supreme Soviet brought any changes in the top leadership. Worse still, contrary to the Party Statutes adopted at the 27th Party Congress, and despite Gorbachev's announcement that there would be a special CC plenary session on cadre questions in the autumn, the session was continually postponed.

When finally held, on 27–28 January 1987, the session achieved considerably less than Gorbachev must have hoped for. Dinmukhamed Kunaev, already relieved as Party boss in Kazakhstan in December, lost his Politburo seat, and Alexander Yakovlev, the CC Secretary for Propaganda, was promoted to candidate membership. But this was a far cry from what most observers had expected.

Not only did Ukrainian Party chief Vladimir Shcherbitskiy (a Brezhnev protégé who had been under attack for more than a year) survive, there were also no promotions for Gorbachev's key allies, such as Boris Yeltsin.

In his keynote address to the plenum, Gorbachev again called for a revolution from both the top and the bottom. Yet, for all his assurances that he strongly supported the strengthening of 'democratic centralism', thus implicitly confirming both the existing system and the primacy of the Party (though suggesting the introduction of secret ballots and more genuine choice in some elections), the final resolution endorsed few of his proposals. It seemed clear that Gorbachev still did not control the CC, and that to attain such control he will have to change its composition as soon as possible. His proposal to hold an extraordinary Party Conference in 1988 (the first since 1941) therefore made considerable sense from his point of view, but it also demonstrated his weakness and gave a hint as to his timetable for consolidating his authority. At the 27th Party Congress in March 1986 there had been obvious signs of disagreement on cadre policy within Gorbachev's heterogeneous power coalition. The January 1987 CC plenum and its results made it plain that disagreements still existed – the major impression was that Gorbachev had been reined in by a collective leadership determined to keep his drive under control.

Changes in the Soviet Political Hierarchy (March 1986 to January 1987)

	Sacked	Appointed	Present Total
Politburo:			
Full members	1	0	11
Candidate members	0	1	8
Central Committee Secretaries	1	2	12
Heads of Department in Central Committee	2	2	21
Republic First Secretaries	1	1	14
Oblast Committee First Secretaries	17	16	157
USSR Council of Ministers:			
Chairmen	0	0	1
First Deputy Chairmen	1	0	3
Deputy Chairmen	2	3	9
Ministers and Chairmen of State Committees	22	23	96
Chairmen of Republic Councils of Ministers	2	2	15

The personnel changes which had taken place at the lower echelons of power during the year (see table) were for the most part a mixed blessing for Gorbachev. Either they had been made in order to discipline particular republics or else they benefited the Prime Minister Nikolai Ryzhkov as much as, or more than, Gorbachev himself. Personnel in some of the Republics which had been criticized escaped virtually unscathed, but in several of the Asiatic Republics there were brutal purges – one minister in Uzbekhistan even being shot. That these moves were not unopposed was vividly demonstrated when Kunaev's dismissal led to violent street demonstrations in Alma Ata. Growing tension between Moscow and the Brezhnevite periphery, fuelled by a drive to Russify the provincial leaderships, together with the broadening of client support by Ryzhkov and Central Committee Secretary Yegor Ligachev, combined with Gorbachev's problems to create a potentially dangerous mixture. Unless Gorbachev can change this trend swiftly, his options both for imposing his own authority and for achieving his agenda might narrow significantly.

Opening Bids on the International Scene

Gorbachev's mounting domestic problems made success in the foreign arena all the more imperative. Not surprisingly, therefore, Soviet diplomacy was marked by unprecedented activism. Arms-control proposals multiplied throughout the year, and a major diplomatic offensive towards Asia was launched. In his July speech in Vladivostok, Gorbachev not only revived the old Soviet plea for an Asian security system but stretched out his hand to China, offering to reduce Soviet troops in Mongolia and Afghanistan and even expressing willingness to discuss some minor territorial concessions. The replacement of Babrak Karmal in Afghanistan was followed at the turn of the year by new promises, including the declaration of a unilateral cease-fire and vague allusions to a speeded-up timetable for a Soviet withdrawal. Relations with North Korea were consolidated during the year, and initiatives in the South Pacific were pursued. And, despite clear differences on Bloc policies in the Kremlin, the economic and technological integration of COMECON was accelerated.

Impressive as this list is, there was something hollow to these moves. They were opening bids, not clear indications of a true willingness to settle outstanding problems. Soviet arms-control proposals invariably preserved what the West perceived as the key problem: Soviet military advantages which ensured for the USSR not only strategic offensive options but also the capability eventually to transform military power and existing imbalances into political influence. The 'new political thinking' had more than a touch of *déjà vu* about it, a throwback to the 1950s and 1960s, reminding the world of the sterile debate about 'general and complete disarmament' and the Rapacki plan. The announced withdrawal of some 6,000 troops from Afghanistan proved to be a sham when it was carried out, and moves to encourage a political settlement there were not accompanied by any sign that Moscow might indeed be looking for compromise; they simply added political pressure to the firm military pressure already being applied to the resistance. The offers made to China would continue to be unrewarded so long as the USSR continued to ignore the most important of China's three conditions: the ending of Soviet support for Vietnam's attempt to bring Kampuchea under its control. And Gorbachev's exhortations characterizing Europe as 'our common house' could only sound cynical after the Chernobyl reactor catastrophe and the Soviet refusal, for crucial weeks, to provide the information on the disaster that its neighbours desperately needed, and to provide compensation for the resulting damage in other countries.

Thus, while undoubtedly there was some movement in Soviet foreign policy, that movement has yet to lead very far. The old hope that its impact on Western public opinion would compel the US and Western Europe either to settle on Soviet terms or risk the delicate fabric of NATO security consensus was still the dominant aspect of Soviet diplomacy. There was a clear determination to win the high ground in arms control and the East–West dialogue, but it was not completely successful. There has been a change in style and dynamism, but not yet one in essence. It cannot be ruled out that Gorbachev may deliberately not have aimed higher; stronger efforts to adjust and reinvigorate Soviet foreign policy may have met the same resistance that greeted his domestic programmes. The new General Secretary still has to prove himself both willing and able to follow up his opening bids with a readiness to pay the price necessary to reach agreements.

The Economic Front

The Soviet economy showed some signs of recovery during 1986. However, these were offset by the accident at Chernobyl and its longer-term repercussions. Furthermore, the adjustments made to Soviet economic strategy might not work as expected and might eventually even undermine prospects for improvement.

The economic results announced at the end of the year sounded optimistic: national income had grown by 4.1% (3.1% in 1985), industrial production by 4.9% (3.9%) and agricultural production by a stunning 5.1% (1.4%). Oil production reversed its decline and rose to 615 million tonnes (595), while coal production reached 751 million tonnes (726), natural gas 686 billion megawatts (643), and electricity 1,599 billion kilowatts (1,545).

But there were many factors that seriously clouded the picture. Some of the rosy data (for example, the national income figure) smelled of imaginative bookkeeping. Growth rates had generally shown a tendency to flatten out during the second half of the year. Capital investments, crucial for future growth, were seriously behind targets, by as much as 50% in fact. And quality continued to be deficient, with only 15% of total industrial production reaching the quality levels attained in the West.

Worst of all, there were factors that rendered the new Five Year Plan target obsolete from the outset. The Chernobyl disaster, the worst nuclear accident ever, was a major blow. Fifteen percent of Soviet nuclear power-generating capacity was knocked out for a considerable time, and the gigantic task of decontamination, cleaning up and caring for the refugees and evacuees (numbers peaked in early May, at roughly half a million people) cost billions of roubles. The resultant rerouting of massive amounts of materiel, personnel and transport resources seriously strained the inefficient planning machinery and wrought havoc with economic growth targets. Scores of projects had to be delayed so as to siphon off resources to the disaster area, and the indirect repercussions (including the toll in dead and cancer victims, whose ultimate magnitude remains difficult to gauge) will be felt for years to come. The combined impact of the Chernobyl disaster, a particularly dry summer and the coldest winter in decades nullified the production gains in energy resources. The USSR, which depends heavily for its future growth on energy and new technology, was hit by Chernobyl in a most vulnerable place.

Gorbachev's economic strategy was also affected by the need to adjust to political pressure. Though the official defence budget remained 4.6% of total government expenditures, its real growth rate was doubled from 3% to 6%. There was a real danger that Gorbachev's long-term national security strategy, already jeopardized by economic and foreign-policy realities, would have its very basis questioned and remain a dead letter, should resistance by the conservative bureaucrats combine with inbred preferences for traditional answers and narrowed horizons. Gorbachev's remarks to the January CC plenum contained suggestions to this effect.

Prospects

The Soviet Union has entered a critical period that gives rise to some fundamental questions. First, change has definitely begun, but can it be sustained against mounting resistance, and where will it ultimately lead? The next few years will tell, both for Gorbachev and for the USSR. Time is not necessarily on the side of either of them. If Gorbachev cannot crack the domestic opposition fairly quickly, he risks being reined in even further by a collective leadership marked by heterogeneous interests and latent rivalries. Any chance of steering the Soviet Union onto a new and more promising course would probably then fade. And in foreign policy, the window available for seeking a *modus vivendi* with President Reagan is rapidly closing. Once the US election campaign starts to cast its long shadow, progress will become much more difficult, if it is indeed possible at all.

Secondly, although Gorbachev is a man with a vision, would the realization of that vision truly correspond to the West's best interests? Gorbachev has not yet shown any interest in agreement *per se*; his objective is rather to preserve the Soviet Union's position as a super-power with global capabilities and ambitions. If the reforms he has suggested are effective, and the USSR becomes more efficient, then it may become more interested in, and more capable of, a constructive dialogue with the West – but there is no *inherent* link here. The breathing spell Gorbachev is seeking may offer the West an opportunity; but it may just as easily be meant to assist a Soviet strategy of *reculer pour mieux sauter.* A Soviet Union on the decline, frustrated by the constraints of other powers and increasingly vulnerable is not a pleasant prospect (in all probability it is a highly dangerous one), but neither is a reinvigorated USSR, both capable of and intent on challenging the West. Gorbachev will have to prove that a third choice exists.

Finally, Gorbachev has run into serious problems. Can – or should – the West strengthen his hand? The old picture of the 'hawks' and 'doves' competing in the Kremlin is a cliché, and may be a dangerous one under present circumstances. Gorbachev has clearly used foreign policy as a tool to strengthen his position and authority at home, and he is a man deeply in need of success. But, so long as his success does not result in a fair and balanced agreement, prudence seems imperative. The USSR cannot be squeezed into agreements, nor compelled to sacrifice its interests. To move carelessly in trying to seize the alleged opportunity of the hour would therefore not only be premature and dangerous but would also misread the situation in the Kremlin. Above all, it would send the Soviet leadership the wrong message. Cheap gains would not strengthen Gorbachev's hand; they would simply convince the more recalcitrant men in the Soviet leadership that difficult choices can be avoided and traditional answers may still suffice. Before trying to help Gorbachev consolidate his authority, the West should be certain that, if he does attain full power, he can be expected to use it in the interests of international interdependence and stability.

Subscriptions to IISS Publications

Three alternative forms of annual subscription are available, covering publications which are issued during the year from the subscription commencement date

COMBINED
1 *The Military Balance*
1 *Strategic Survey*
12–15 Adelphi Papers
6 issues of *Survival.*
Subscription price:
£65.00 ($US 97.00)

SPECIAL
1 *The Military Balance*
1 *Strategic Survey*
12–15 Adelphi Papers
Subscription price:
£60.00 ($US 90.00)

SURVIVAL
6 issues of *Survival*.
Subscription price:
£20.00 ($US 30.00)

Published during year from subscription commencement date to be entered below.

If you would like to take out one of these subscriptions, please fill in and mail the card below.

SUBSCRIPTION CARD

Please enter subscription ticked below for 1 year from ____________________ (date)

Combined ☐ £65.00 ($US 97.00)
Special ☐ £60.00 ($US 90.00)
Survival ☐ £20.00 ($US 30.00)

Prices applicable to subscriptions commencing between 1 January and 31 December 1987 only)

Payment: ☐ Payment enclosed
☐ Bill me

I am a **new subscriber** ☐

I already have: Combined subscription ☐
Special subscription ☐
Survival subscription ☐

(please enclose IISS mailing label)

SE T E AND RESS RLY

Name ____________________

Address ____________________

City ____________________ Country ____________________

THE INTERNATIONAL INSTITUTE FOR STRATEGIC STUDIES
23 TAVISTOCK STREET
LONDON WC2E 7NQ
ENGLAND

Europe

SECURITY UNCERTAINTIES IN NATO COUNTRIES

The political and military establishments of the European members of NATO have had another worrying year. Their loyalties have been tested by the US Administration's commitment to a Strategic Defense Initiative whose implications are uncertain. They have had to cope with a Reykjavik summit that, if some of the suggestions made there had been implemented, would have resulted in the *de facto* abandonment of NATO's strategy of Flexible Response, at least as it is currently understood. They are concerned that, given the Soviet strength in conventional forces, any downgrading of the role of nuclear weapons would leave their security impaired. To add to these problems, there has been an increase in terrorist violence and difficulties in co-ordinating action against it.

Western Europe has also seen the breakdown of the domestic consensus on security policy that held until a very few years ago. A number of Socialist opposition parties have made significant policy changes on defence, with particular emphasis on the need to renounce the early use of nuclear weapons. In Britain the main opposition party has gone further, declaring its opposition to NATO's nuclear strategy and its intention, when in government, to abandon the British nuclear deterrent and require the removal of all US nuclear bases from the country.

Arms Control and Strategy

The persistent concern of the European NATO leaders is with the maintenance of strategic stability. They need to work out how to cope with the growth of Soviet military power when defence budgets are static at best, the numbers of men in uniform are falling, and opinion in the United States, notably in Congress, is showing signs of an impatience with European security problems that may lead to renewed pressure to withdraw some US troops.

More immediate, however, were European worries over the American attitude towards arms control and NATO strategy at the Reykjavik summit. Both Chancellor Kohl and Prime Minister Thatcher clearly saw a need to discuss the issues soon after the summit and visited Washington for talks with President Reagan, on 20 October and 14 November respectively. Both took care in public to stress the unity of the Alliance behind the US position at Reykjavik and to praise the progress made, though in private they expressed their concerns about some proposals made at the summit. Now the European allies face with mixed feelings the Soviet proposal made on

28 February 1987 for eliminating intermediate-range nuclear forces (INF), notwithstanding the fact that the 'zero option' was originally proposed by the West.

The December 1979 decision by NATO to deploy INF in Europe – or to bargain deployment for the removal of Soviet SS-20 missiles – was an attempt to restore credibility to the US nuclear guarantee to Europe. It has been the subject of debate whether this credibility was meant to be enhanced by the political symbolism of the deployment, by its provision of nuclear options to match those open to the Soviet Union, or by a combination of both. What is clear, however, is that, although the NATO military (and the French government) profoundly dislike the zero option, allied governments are likely to be ready to go along with an INF agreement on this basis, provided it also deals satisfactorily with the shorter-range nuclear missiles and verification. The European wish for arms control is too strong for any other reaction, and this would appear to be true of the Reagan Administration as well.

European governments continue to have doubts about the Strategic Defense Initiative (SDI) and its consequences for the Alliance. They are still unclear (and they are not alone in this) about what the US sees as emerging from the research currently under way and what the deployment of strategic defences would mean for NATO strategy in the long term. For the moment, however, those nations which have agreed to participate in the research programme at governmental level – Britain, Germany and Italy (France reserves its position on the programme while not preventing French companies from taking part) – have clearly concluded that, if they cannot stop it, they would be unwise not to try to share in the technological development and commercial spin-off possibilities it is likely to produce.

Yet there are signs of increased nervousness about reports that the US may adopt a broad interpretation of the 1972 Anti-Ballistic Missile (ABM) Treaty in order to make early progress on the testing and deployment of SDI systems. European governments tend to see the ABM Treaty as the only remaining arms-limitation agreement of the 1970s, and are particularly reluctant to see it eroded. The US sent Paul Nitze, special Adviser on Arms Control and Richard Perle, then Assistant Secretary of Defense, to European capitals in February 1987 in an effort to allay European fears. They emphasized that no decision had yet been made on adopting a broad interpretation, and assured the allies that they would be consulted before any decision was taken – but, no doubt, they also sounded out allied reactions to the broad interpretation.

Socialist Party Positions

Developments over the past year in several European Socialist parties have demonstrated a resurgence of anti-nuclear feeling and idealistic aspirations for effective but non-provocative conventional defences. There are indications that these shifts in policy are deep-rooted and unlikely, as in the past, to be readily abandoned on accession to power.

Of the two major parties that have swung leftwards on security issues, the British Labour Party offers the more serious challenge to the traditional security consensus, because of its sweeping rejection of the concept of nuclear deterrence. The West German Social Democratic Party (SPD) appears prepared to accept continued reliance on the US nuclear guarantee, but its ideas would nonetheless pose problems for NATO's conventional doctrine, given its view that the present defences are provocative. The concurrence of these two parties' deviations from present NATO policy particularly worries the NATO alliance, because of the unique importance of Britain as a reinforcement base for the United States and of West Germany as a front-line state and the strongest conventional power in Europe.

Nuclear Issues

The British Labour Party has gone well beyond the simple opposition to new NATO missile deployments expressed by various European parties in the early 1980s. It now renounces the entire American nuclear umbrella and proposes to require the withdrawal of all US nuclear forces from Britain, including the missile submarines based in Scotland and the nuclear weapons for the F-111 strike aircraft in England. Initially Labour talked of a one-year deadline for the withdrawal, but in the face of widespread NATO unease at the idea it has since blurred its position by acknowledging a need for prior consultation (though not necessarily agreement) with allies. In March 1987, taking into account the renewed possibility of an agreement on INF, it further modified its position on cruise missiles in Britain by saying that they could remain while significant negotiations concerning their removal were under way between the super-powers.

The party is also committed to abandoning the British independent nuclear deterrent: it would cancel the planned acquisition of *Trident* submarine-launched ballistic missile systems, decommission the existing *Polaris* systems, and also dismantle all British tactical nuclear weapons in Britain and Germany. Labour advocates a NATO policy of no first use, withdrawal of all nuclear weapons from Europe and then a purely conventional NATO strategy. (How non-nuclear British forces would take part in the defence of the central front if the NATO strategy remains nuclear has not been explained.)

Labour has also endorsed joint proposals by the West German SPD and the East German Socialist Unity (Communist) Party (SED) for a nuclear- and chemical-weapons-free zone along the East–West German border. In its joint recommendations with the SPD in November, it also advocated a freeze on nuclear testing and the abandoning of European support for SDI research. The party maintains that its non-nuclear defence policy would be compatible with continued support for NATO, but American officials have questioned whether US forces could remain in Europe without nuclear protection. There are real fears that rejection of the NATO doctrine of Flexible Response by a key member country could well lead to an unravelling of the consensus that holds the Alliance together.

For its part the SPD, at its party convention in Nuremberg in August, repeated its opposition to INF (by implication anywhere in Europe, and not just in West Germany). At the same time the convention called for the removal of Soviet shorter-range missiles 'deployed as a countermeasure in the GDR and Czechoslovakia'. The SPD also endorses a policy of no first use of nuclear weapons and seeks the 'gradual creation of a nuclear-weapon-free Europe'. But the moderates secured party agreement that any removal of NATO nuclear weapons would be a matter for Alliance consultation, rather than unilateral West German decision. The most dramatic element of the SPD's approach, however, was its inter-party negotiations with the East German SED for a zone free of nuclear and chemical weapons, 150 kilometres wide, astride the East–West German border. It broke new ground in startling fashion by issuing a joint appeal with a Warsaw Pact government party, and had no qualms about treating East Germany as sovereign in such security decisions.

In the Netherlands, despite the commitment by the centre-right government to implement cruise missile deployment, the Dutch Labour Party has been surprisingly quiet on the INF issue in the past year. Officially the party now seeks 'minimum deterrence', with a shift to greater reliance on conventional deterrence and reduced dependence on potential nuclear escalation. It seeks to limit the Netherlands' nuclear role to the operation of one weapons system – and would prefer that system not to be the cruise missile, so as to be free of the INF problem (a dilemma from which Mr Gorbachev may now have released them).

In Denmark a long-standing national policy prevents the presence of nuclear weapons on Danish soil in peacetime, so there has been no question of INF deployment there. At the height of the INF controversy the Social Democrats did force the government to oppose INF deployment in other countries and to vote against NATO funding to support this deployment directly or indirectly. They now advocate barring nuclear weapons from Denmark altogether, including in time of crisis or war.

Alternative Defence Strategies

All the socialist parties in NATO member states pledge support for NATO (though there is a left-wing group within the British Labour Party that is strongly against even this). German left-wing proposals – like that of the Saarland Premier Oskar Lafontaine, urging the Federal Republic to imitate France and leave the NATO integrated military organization – are endorsed neither by the SPD nor by any other major socialist parties. Nonetheless, most parties (though not in France) have come to question the wisdom of NATO doctrine, which many view as inconsistent with the idea of a 'defensive' alliance. This has led to a plethora of proposals for the adoption of 'non-offensive', 'non-provocative', or 'popular' defence strategies, so that the USSR and its East European allies would have no cause to see NATO as threatening. After various ideas on alternative defence had been pioneered by the Scandinavian and Benelux socialist parties in joint meetings beginning in 1981, the SPD's left wing embraced them wholeheartedly.

It was explicit in this approach that NATO's assessment of a Soviet-bloc numerical superiority in heavy weapons in the central region should be discounted, on grounds of the technological inferiority of Soviet-bloc weapons, undermanning of Soviet units, low levels of education and training (especially in electronics), a dearth of initiative in the Soviet forces at junior levels of command, and the unreliability of East European troops. Explicit also was the view that Moscow would not initiate hostilities unless provoked by the West. Among examples of Western provocation cited were the *Pershing* II and cruise missile deployments, the US Army *Airland Battle* doctrine, the NATO concept of Follow-on Forces Attack (FOFA), and West Germany's possession of twice the number of tanks that Hitler had when he attacked the Soviet Union.

Hence the SPD's Nuremberg platform calls for armed forces that are 'structurally incapable of aggression'. But the party also advocates, in wording inserted by the moderates, 'reform of the armed forces' to 'develop the capacity for forward defence, which promotes stability'. This would include strengthening 'anti-tank defence, interdiction and air defence' and, most surprisingly, erecting barriers on the North German Plain, despite long-standing West German opposition to them. Air strikes against such East European targets as airfields, bridges and railway junctions are not explicitly ruled out, but the strong implication is that no major counter-attack by NATO ground forces into Warsaw Pact territory would be countenanced.

The most serious rejection of NATO policies at the SPD's Nuremberg convention was in the proposal to terminate the 1982 Host Nation Support agreement, under which equipment for six US reinforcement divisions is pre-positioned in the Federal Republic, and West Germany promises to mobilize 100,000 reservists to provide logistic support for these divisions. The reason given for this rejection is that the US might use these forces for national purposes outside the NATO area without Bonn's consent. This resolution was a last-minute surprise successfully sprung on the convention by the left, and party leader Johannes Rau omitted this position from his own platform for the general election in January 1987.

The SPD argues that its reforms would not require additional expenditure, but rather would cut defence spending from 20% of the federal budget to the figure 'under the last SPD-led government' (i.e. 18.4%). The party also rejects the recent extension of conscript service to 18 months, but would continue conscription, letting

Bundeswehr numbers decline as the pool of draft-age men shrinks in the 1990s. Without offering precise figures it says that 'sizeable elements of the armed forces must be skeletonized', with the gap between peace and war strengths offset by better use of reservists. It is, however, significant that the parliamentary party has followed a more centrist line than the party convention, voting in recent years for a variety of weapon systems, regardless of their lack of unambiguously defensive characteristics.

The British Labour Party defence statement issued in December 1986 charged the Thatcher government with building nuclear systems at the expense of conventional capabilities, and said that Labour would devote the money saved by cancelling *Trident* to increased conventional spending. Like the SPD, Labour does not face the fact that conventional forces are more expensive than nuclear forces and has indeed earlier said that its aim over time would be to reduce defence spending to the level of its major European allies.

Labour's 'restructuring' of the armed forces would reinstate the commitment to a 50-warship Navy, build the European Fighter Aircraft, modify the *Tornado* interdictor/strike variant for a battlefield role and 'restore the standards of equipment and training of the British Army in Germany'. It does not contemplate a reintroduction of conscription in order to strengthen conventional forces. And, in a phrase that makes many West Germans uneasy, it calls for 'defence in depth' in Germany rather than forward defence.

The Danish Social Democrats, too, share the goal of a 'purely defensive' defence. In a policy paper circulated in the second half of 1986 the party called for Danish warships to be replaced by missiles on shore and (in surveillance missions) by fishing boats. It also sought the removal of one of the three Danish brigades from Schleswig-Holstein and implied that Denmark's F-16 aircraft should be made incapable of striking air bases in Eastern Europe. It also strongly implied that Denmark should not commit its own forces to battle as soon as any Warsaw Pact attack was launched on West Germany but hold aloof until the invasion reached Danish soil. Both Britain and the Netherlands, the allies committed to sending troops to Denmark's defence, have taken the suggestion seriously enough to warn Copenhagen informally that reinforcement would be impossible in such circumstances.

Not all Socialist parties were moving against the grain, though. Unexpectedly, Spanish socialists are turning into solid supporters of NATO. Spain, like France, is a member of the Alliance but not of NATO's integrated military structure. Unlike France, however, it is a member of the Defence Planning Committee and the Military Committee and sends observers to the Nuclear Planning Group. Since winning popular approval for Alliance membership in a referendum in March 1986, Premier Felipe González has stressed co-operation with NATO. A government memorandum leaked in October spelled this out a little further, saying that Spain was willing to contribute to the common defence and to co-ordinate military tasks, such as naval patrolling, with NATO.

During the referendum campaign, however, González had made his support for continuing NATO membership conditional on a substantial reduction in the 12,500 US troops stationed in Spain. Talks on extending US base rights past their May 1988 expiry had reached an impasse by the end of 1986, with Spanish officials insisting on US withdrawal from the air base at Torrejón, near Madrid. Thousands of Spaniards staged on angry protest against the US military presence on the eve of Secretary Weinberger's visit in March 1987, which was aimed at heading off the Spanish threat to force a complete US pull-out unless the government's demands were met. The pressure on González was compounded by a wave of demonstrations protesting at the government's economic and social policies, leaving the Prime Minister's popularity badly damaged less than a year after his impressive general election victory in June 1986.

The French Socialists are also continuing to support the policy, initiated when they were in power in the early 1980s, of intensifying Franco-German military co-ordination and holding joint exercises in Germany, thus edging gently back from de Gaulle's withdrawal from the NATO military structure. And Italy, which had a Socialist-led coalition government for some 3½ years until March 1987, supported NATO policy throughout; indeed the deployment of cruise missiles at Comiso was the only one of these deployments in the Alliance to take place without major demonstrations.

Inter-party Co-ordination

Some effort is being made by moderate Socialists to provide a common denominator for the various parties' security policies – and to pull them more towards the centre. At the third meeting of NATO Socialist parties in Oslo in September 1986 the main thrust of the communiqué was that the Europeans should unite their security policies to gain influence within NATO 'especially in the field of arms control' and East–West relations, and that exploration of the possibilities of nuclear- and chemical-weapons-free zones and no first use of both conventional and nuclear weapons should be undertaken. But in terms of concrete defence policy the communiqué did not challenge NATO doctrine in nuclear or conventional terms. In the nuclear field it deliberately exercised restraint so as not to burden super-power negotiations in a pre-summit period, demanding no unilateral disarmament but only 'minimum nuclear deterrence' as a step towards 'elimination of nuclear weapons'. On conventional forces it called with careful ambiguity for 'a conventional balance at the lowest possible level, including the elimination of offensive capabilities'.

Both the British Labour Party and the Danish Social Democrats agreed to these formulations, and moderate West German Social Democrats were instrumental in engineering their adoption. If present trends continue, however, a further leftward drift of socialist parties is possible over the long run. It is improbable that the Labour Party would abandon its current policy if elected to office, for many of the younger leaders are more radical than their seniors, and in any case its leader, Neil Kinnock, has a personal commitment to nuclear pacifism. These radical views, however, do not command majority support among the British people.

In the case of West Germany, nuclear pacifism has probably taken deeper root, both in the SPD and in public thinking, than it did in the old *'Atomtod'* days that preceded the party's famous pragmatic turn at the Bad Godesberg Convention in 1959. The Social Democrats are now battling with the more radical Greens for votes. And the French Socialists no longer exert much influence on the SPD, for party bilateral defence consultations have virtually ceased.

Effect on Key Governments

Socialist defence policies thus often represent a significant challenge to the prevailing NATO consensus. However, Alliance leaders – civilian and military – have categorically rejected the viability of non-nuclear doctrine and 'defensive defence' as a means of countering the Soviet threat. For the time being, therefore, the principal governments in NATO seem unmoved by the socialist thinking, and political events in 1986 (notably the SPD's failure to attract significant support in the West German general election) have tended to reinforce the impression that fundamental change is not imminent. Britain's Conservative government under Mrs Thatcher is quite determined to go ahead with the modernization of its nuclear deterrent and to support both NATO nuclear policy and multilateral disarmament. But attention will be focused on Britain throughout the Alliance, since the debate over NATO's security policy is certain to play an important role in the forthcoming election and is likely to remain politically divisive, whatever the outcome.

In France, Prime Minister Jacques Chirac and Socialist President François Mitterrand established a wary *modus vivendi* following the election of a conservative majority in the National Assembly in March. The principal controversies that dogged France's uneasy political 'cohabitation' concerned the privatization of industries, banks and insurance companies nationalized by the Socialists in 1982, election law reform and, at end of the year, a fierce national dispute over the government's proposals for university education which brought hundreds of thousands onto the streets in protest and led a shaken Chirac to withdraw the legislation.

The broad French consensus over the main elements of security policy, however, assured that differences over defence questions were primarily ones of emphasis. The five-year defence programme for 1987–91, announced in November, called for a 7% increase in defence spending for 1987 and funding for a number of new procurement programmes, including France's first nuclear-powered aircraft carrier (the *Richelieu*), a new main battle tank, a fighter (expected to be the Dassault *Rafale*) and airborne early-warning aircraft (the American AWACS). The nuclear programme represented a victory for President Mitterrand in the priority it attached to modernizing the nuclear submarine-launched ballistic missiles (beginning with the M-4 six-warhead missile, to be followed by the M-5 with 9–10 warheads) while deferring replacement of the land-based missiles until the mid-1990s, leaving open a decision whether those missiles should be mobile (an idea advocated by Chirac). The defence plan was also seen as a rebuff to Chirac's suggestion that French shorter-range nuclear missiles (currently *Pluton*, to be replaced by the somewhat longer-range *Hadès*) should be used as a tactical system – rather than serving as a 'pre-strategic' last warning before strategic systems come into play. The plan also indicated that the government was prepared to begin production of chemical munitions.

In Norway, the Conservative government lost a crucial vote in the *Storting* (Parliament) in April on an austerity budget package (to offset the effects of the sharp decline in revenues caused by falling oil prices) and was succeeded by a minority coalition under Labour's Gro Harlem Brundtland. The new government gained approval for its own budget measure, involving sharp cuts in public spending, and continued to maintain a working, if tenuous control, of the *Storting* for the rest of the year. In security policy the new government represented a slight shift to the left, particularly in its more vocal criticism of SDI and its readiness to engage in discussion of a Nordic nuclear-free zone (though there were suggestions that this was more for presentational purposes than a reflection of real intent). It did, however, promise to sustain defence spending increases, proposing a budget with 3% real growth, slightly below the 3.5% advocated by its predecessor.

The Problem of Terrorism

Another security threat causing difficulties for the European countries was terrorism. At the beginning of 1986 the United States stepped up its drive against the Gaddafi regime, which it saw as a key agent for state-supported terrorism. Although EEC Foreign Ministers were prepared to agree on a general arms embargo on 'any country implicated in supporting terrorism', they refused to single out Libya as the US wished. In the wake of the American air strikes on Libya on 15 April, however, the Foreign Ministers decided to slash the size of Libyan diplomatic missions in Community capitals – an action that met with US approval. Later in April, Community Ministers of the Interior and Justice (the 'Trevi' group) agreed to meet to discuss counter-terrorism and related matters at least twice a year at ministerial level – the session at which this was decided being only the third such meeting since 1976.

The group next met in London on 25 September, after a bomb attack on the Paris town hall on 8 September heralded a wave of explosions in the French capital which left ten dead and hundreds injured – apparently an effort by a Lebanese-based group to force France to release Georges Ibrahim Abdallah, a leading terrorist who was awaiting trial. Faced with this and similar terrorist demands, the meeting agreed to step up the fight against terrorism and agreed, *inter alia*, to examine visa procedures with a view to excluding known terrorists from Community states, to move towards more effective extradition procedures, and to ensure that member states would make no concessions under duress to terrorists or their sponsors.

These eminently sensible procedures unfortunately collapsed within months. There was a continuing series of terrorist incidents in Europe: in Bonn the chief of the West German Foreign Ministry political department was shot dead by the Red Army Faction on 10 October; bombings continued in France; and in Britain a Syrian, Nezar Hindawi, was sentenced to 45 years in prison for trying to plant a bomb on an El Al aircraft with 380 people aboard at Heathrow Airport in April. The British government had evidence that the Syrian Embassy was involved and therefore broke off diplomatic relations on 24 October, urging its Community partners to follow suit. All except Greece eventually agreed to take some punitive action against Syria, but several governments resented the pressure that Britain had exerted. The Trevi efforts to develop a united front to deal with terrorism meanwhile continued with meetings in November and December. The December meeting agreed a political analysis of the terrorist threat, but again the Greek government refused to sign; no reasons were given, but it was widely surmised that Greece objected to certain states, such as Syria, Iran and Libya, being identified as involved in sponsoring terrorism.

The credibility of further international efforts was then sharply undercut by the revelations of the secret US arms-for-hostages dealings with Iran. The dilemmas posed for governments by the surge of kidnappings in Beirut became more acute in January 1987, when two West German businessmen were seized, presumably as hostages to bargain for the release of two Arabs held on terrorist charges in Germany. France faces a similar dilemma over whether to trade an imprisoned Iranian in return for the release of French hostages being held in Beirut by pro-Iranian groups.

There were signs that both Bonn and Paris were considering some kind of deal, but any French effort to find a way out was set back when, despite the prosecution request for a lenient sentence, a special court sentenced Abdallah to life imprisonment for his involvement in the murders of an American and an Israeli. Should the French and German governments move towards exchanges of prisoners for hostages, it would in the short run be very awkward for Britain, which would come under intense pressure to negotiate for the release of Terry Waite (the Archbishop of Canterbury's envoy, who went to Beirut to try to secure the release of foreign hostages but was himself seized in January 1987). In the long run, and even in the light of the revelations about the American machinations, it is difficult to see how deals by governments could do anything but encourage further kidnappings and demands. The world would then become a more, not less, dangerous place in which to live.

Another Difficult Year Ahead

The re-election of Chancellor Kohl in West Germany in January 1987 and the poor showing of the Labour Party in the run-up to a possible election in Britain appear to have removed the immediate threat to the NATO consensus raised by the German and British socialist parties. But there is no sign that it is a threat that will be easily surmounted in the long run. Socialist defence policies look likely to remain a significant obstacle to smooth relations between Europe and the United States. Tensions will also continue to develop around trade and protectionist questions; not only are agricultural policies a sore point, but industries such as aircraft manufacture are now the subject of intense transatlantic accusations. And, even if a US–Soviet INF arms-control agreement is reached after full consultation with the European governments, the United States' pursuit of its SDI goals, coupled with the possibility of a broad re-interpretation of the ABM Treaty, remain as constant irritants in the Alliance. If 1986 was a worrying year for NATO leaders, 1987 appears unlikely to offer smoother sailing.

EASTERN EUROPE: STILL ADRIFT

Two years after Mikhail Gorbachev took command in the Kremlin, the regimes of Eastern Europe are still stuck with a familiar dilemma: how to reconcile the pressures on the home front for reform and the assertion of national identity with the pressure from the Soviet Union to toe the ideological line. If anything, the dilemma has worsened. On the one hand, it is still not clear how Gorbachev intends to manage his allies in COMECON and the Warsaw Pact, nor precisely which line they might be expected to toe – nor, indeed, how far his own internal reforms will go. On the other hand, the new, more active Soviet diplomacy towards Western Europe has only served to encourage what might best be called the creeping Europeanization of Eastern Europe, as the societies there grow increasingly sophisticated, both economically and politically.

In the early part of 1986 the new Soviet leader took the trouble to acquaint himself at first hand with the problems of the region by visiting several East European countries in rapid succession. He also seemed to be embarking on a Soviet version of a 'Year of Europe' that would have given these countries a special role in Moscow's policy towards Western Europe. However, his attention soon wandered.

The battle to consolidate his power at home, and the distracting ups and downs of Soviet–American relations still commanded most of his attention.

As a result, the political signals from Moscow to Eastern Europe were at best conflicting, at worst thoroughly confusing. Gorbachev himself avoided any public reference to the Brezhnev Doctrine – the USSR's traditional assertion of a right to intervene in Eastern Europe to uphold ideological orthodoxy – and seemed to hint at greater toleration of diversity, if only in the debate about economic reform which he unleashed in the Soviet press. However, the message hammered home repeatedly by Soviet ministers and officials was of the need for greater economic cohesion and greater East European attention to the needs of the USSR.

As heir to an empire that had been in dire economic and political trouble in the early 1980s, particularly (but not solely) as a result of the Polish crisis, Gorbachev might have been expected to set out his thoughts on the management of the Warsaw Pact at the Soviet Party Congress in late February and early March 1986. In the event, he dropped few clues. Although he spoke of paying more attention to the views of others and affirmed that there were indeed different paths to socialism, his remarks seemed to be addressed more to China, whom he has wooed assiduously over the past year, than to his allies inside the Soviet fold.

Yet, as Gorbachev began his second year in power, all the East European regimes (with the partial exception of East Germany) were still worried about their own poor economic performance and anxious about the deep chill that had settled on super-power relations in the previous few years. Officials in the region were clearly looking to the new chief for a lead of some kind. In a year when oil prices outside Eastern Europe plummeted still further, the Chernobyl nuclear reactor explosion spread radioactive contamination over large parts of Eastern (and Western) Europe, and the debate about economic reform continued, Gorbachev's reticence about Soviet–East European relations has been puzzling.

Foreign Policy

Nevertheless, from an East European point of view, Gorbachev has done some things right. His resumption of dialogue with the United States, the new Soviet proposals in arms-control negotiations and his charm offensive towards Western Europe provoked sighs of relief from Warsaw to Sofia; the long chill in East–West relations had been damaging economically – particularly for Hungary, whose economy is closely bound up with East–West trade. Yet the new Soviet diplomatic activity also cut some of the ground from under those smaller Soviet allies, such as East Germany, Hungary and Bulgaria, who in 1984 and 1985 had taken advantage of the uncertainty over the succession in the Kremlin to argue that the smaller European states should play a larger role in the East–West dialogue. If the two super-powers were on reasonable speaking terms again, this reduced the value of East European countries as channels of communication and thus their room for independent manoeuvre.

This was clearly shown in the way Moscow used Eastern Europe as part of its new diplomacy. Soviet officials kept the East European regimes in touch with super-power developments in fairly regular briefings, but there was little evidence that their views were canvassed on matters that directly concerned them before decisions were taken in Moscow. They, like the rest of the world, had to piece together the changing picture of Soviet foreign and domestic policy as best they could.

Where Gorbachev did try to involve his allies in Soviet diplomacy, it was to use them as a diplomatic platform, rather than a source of ideas. He used the East German Party Congress in April and the June meeting of Warsaw Pact leaders in Budapest to launch new arms-control initiatives. The Polish Congress later the same month gave him a rostrum from which to appeal to the 'Europeanness' of Europe – and, by implication, Europe included the Soviet Union but not the United States. East European delegations were chosen to present Warsaw Pact proposals at both the Stockholm CDE conference on security-building measures and the MBFR talks in Vienna, but these did not always seem to have been discussed in detail beforehand. And although there was relief and some satisfaction in Eastern Europe at the USSR's Stockholm concessions over on-site inspection, the decision on them appears to have been taken exclusively in Moscow.

After the public disarray of the preceding two or three years, the re-establishment of Soviet initiative in East–West relations restored a degree of harmony on this issue within the Warsaw Pact. At least in foreign policy, Gorbachev left the impression that he was more aware of his East European allies' concerns than his predecessors, even if these were not necessarily uppermost in his mind. A further example of this was the Soviet attitude to the re-establishment of contacts with China. Both Poland's Gen. Jaruzelski and East Germany's Erich Honecker were given the go-ahead from Moscow to visit China in 1986, notwithstanding the continued Chinese resistance to restoring relations with the USSR. In the past it had always been assumed that the East European countries would have to wait for Sino-Soviet relations to be normalized before taking political advantage of the increasing number and variety of economic contacts they had developed with China.

The formula used during Honecker's visit to Beijing – that there was really no need to re-establish Party-to-Party relations, since these had never formally been broken off – has since been picked up by Soviet commentators in pressing China to consider similar ties with Moscow. Clearly this was another case where East European interests happened to coincide with Soviet policy towards China: Gorbachev was ready to adopt a more flexible and tolerant attitude to relations between China and Eastern Europe than his predecessors – but he still kept Soviet interests very much in mind.

The Economic Picture

In his first two years Gorbachev has restored a sense of leadership to the Warsaw Pact, but not, as yet, a clear sense of direction. On his travels about Eastern Europe he delivered his pep talk, already familiar at home, on the need for economic 'efficiency' and 'intensive' growth, but the response was mixed. In East Germany, by far the most stable and successful of the COMECON economies, his calls for new thinking were met by Honecker with smug satisfaction at a job already well done. The Czech Party leaders, who had most to fear from a determinedly reforming Kremlin, picked up the buzz-words without enthusiasm but have so far not felt impelled to do much to back them up in practical terms. Hungary, the most adventurous of the East European reformers, has felt vindicated in its pursuit of reform, but is increasingly unnerved by the combination of social tensions that reform has brought and by the relatively poor performance of its economy during the first half of the 1980s.

One problem for all the economic managers of Eastern Europe is that, until the Soviet Union settles on some form of blueprint for economic evolution, nobody can be certain of the bounds of acceptable change. Gorbachev himself is said to have told East European economic officials not to rely too heavily on market mechanisms to save them, adding 'you should not think about life-savers but about the ship, and the ship is socialism'. He has also made clear that he supports economic centralization of the right kind. At the same time, however, he has spoken admiringly of some of the more decentralizing experiments with reform, particularly in East German industry and Hungarian agriculture.

While it is not clear how much reform of the economic system Gorbachev is prepared to tolerate, he has made it plain what he expects of his allies in terms of results. The campaign for greater co-ordination within COMECON, begun in earnest in 1985, has been stepped up. And, while Soviet pressure on allies to deliver better-quality goods has opened up COMECON meetings to frank (at times heated) debate, the USSR's economic grip is tightening, even on those economies which have geared themselves to trade with the West: goods that might have been sold for hard currency must now be traded eastwards. The only rays of hope on an otherwise bleak horizon are talk of more East–West contacts on an enterprise-to-enterprise basis, bypassing COMECON's bureaucratic machinery of co-ordinated central planning, and further discussion of currency convertibility in COMECON trade, to avoid the need for bilateral trade balancing.

One of the less promising prospects, however, is that for energy. The East European countries have not benefited as quickly as they would have liked from the drop in world oil prices. Because of the way COMECON calculates the price of oil supplied to them by the USSR – on a five-year (some say now three-year) moving average of world prices – the sharp drop in oil prices in 1985 and 1986 will take some time to work its way through. Nor does Eastern Europe seem likely to receive compensation from the Soviet Union for the economic and other damage caused by the explosion at the Chernobyl nuclear reactor. However, without specifically mentioning that incident, Poland – which suffered most from it (and was also badly affected by river pollution resulting from chemical spills in Czechoslovakia) – has suggested that COMECON should adopt new rules on compensation in cases where serious economic damage is inflicted by one member on another.

A further divisive issue in Soviet–East European relations has been continued Soviet pressure for increased defence spending by Warsaw Pact allies; such pressure is, to say the least, unwelcome at a time of continuing economic difficulty. Romania has long made a public stand against Soviet pressure, and a referendum there in September 1986 produced the expected crushing majority for reducing the official defence budget by 5%. Hungary had already made known that it favoured spending existing defence budgets more wisely, and even East Germany had been stung into asserting that it pulls its weight in the alliance. Only Bulgaria had seemed to respond with anything like alacrity, having increased its defence budget sharply each year since 1984.

Given the Soviet Union's official policy of deep reductions in nuclear weapons, leading to their abolition in fifteen years, there is bound to be continued friction over the issue of defence spending, as greater attention is thereby focused on conventional forces in Europe. At a time of economic stringency in the Soviet Union, Gorbachev is unlikely to let his allies off the burden-sharing hook.

Politics and the Future of Eastern Europe

Gorbachev is fortunate to have presided over the winding down of the Polish crisis, rather than its winding up. All the same, he appears to get on surprisingly well with the Polish leader, Jaruzelski. By choosing to appear in person at the Polish Party Congress, he gave his official stamp of approval to the Jaruzelski formula of half-concessions and limited dialogue. The release of political prisoners from the Solidarity period during the summer was the sort of calculated risk that probably needed a prior Soviet nod – though in the event Jaruzelski gained more from reducing his opponents from martyrs to ordinary mortals than he risked from freeing a group of people who are badly divided amongst themselves and lack any sort of coherent programme. In contrast to the continued persecution of the Turkish minority in Bulgaria and other human-rights abuses in Eastern Europe, Poland's record now looks much better. This prompted the US to lift its remaining economic sanctions on Poland, and as a result political, economic and cultural relations between the two were practically normalized.

Of the other issues which have obtruded onto the agenda of Soviet–East European relations, and which will eventually demand attention from Moscow, the most important is the question of political leadership. Although the succession crisis in Moscow appears to be over, at least for some time, almost all the Communist Parties of Eastern Europe are headed by men in their late sixties or early seventies and can expect to see their leaders change at some point in the next five years or so.

Poland's Gen. Jaruzelski is only 63, but his military connections are still an embarrassment, and the Soviet Union would no doubt prefer to hasten a transfer of power if there were a credible alternative leader. Already people in Hungary and Czechoslovakia have begun to speculate seriously on the likely successors to Kadar and Husak. In East Germany, Honecker appears to have manoeuvred his own choice, Egon Krenz, into a strong position, and there have been persistent rumours that he will retire soon, although preferably after making his much-cancelled trip to West Germany. The successions in Hungary and Romania look like being trickiest of all. In Hungary the Kadarist social contract of relative economic openness in return for political quiet could easily come unstuck in less careful hands. In Romania, where the economy is already in dire straits, Ceaucescu's attempt at socialism in one family, with most of his relatives filling the upper echelons of Party and government, could well provoke some attempt by the country's military to intervene in the battle for power that seems inevitable once he goes.

Gorbachev has made it clear, if only by implication, that he would prefer to have about him a group of technically competent comrades in Eastern Europe to manage the economic changes he seems intent on pursuing. The question is: to what extent will he choose, or indeed be able, to intervene to ensure that his preferred candidate gets the job.

The men who inherit the mantle of Party power in Eastern Europe will have to be able to cope with societies that have grown increasingly sophisticated, not only in their demands for material goods, but also in their political aspirations. The Chernobyl nuclear disaster and its consequences in Eastern Europe brought a sharp surge in both anti-Soviet feeling (Poland had to turn to Sweden for information about the accident, since Moscow gave it precious little in the first few weeks) and concern about the long-term economic and social effects of forced industrialization. In Poland, concern for the wasting environment has provided a focus for much opposition activity in the exhaustion that has followed martial law. Small protest movements have also thrived in East Germany, Hungary and Czechoslovakia, and links have been established with like-minded groups in Western Europe, especially in West Germany and Austria. The ambitious COMECON nuclear energy programme is bound to keep the issue of the environment in sharp focus.

There seems to be increasing concern, too, at what is seen as the militarization of society, especially in East Germany. Despite being initially beheaded, when many of its activists were allowed to leave for the West in 1984, the small unofficial East German peace movement has nonetheless managed to thrive under the auspices of the

Church. In Poland a new group called the Movement for Freedom and Peace came to public attention through its objection to the military oath that pledges fealty to the USSR.

To the extent that these new informal groups pick up concerns already expressed by dissident organizations of longer standing, like Charter 77 in Czechoslovakia and Solidarity in Poland, they will become a matter for deep concern to their governments. They have already begun to forge broader links within their respective societies that could cause irritation, if not a direct threat, to the regimes in power. Joint statements – such as that issued by 122 dissidents from the GDR, Czechoslovakia, Hungary and Poland on the thirtieth anniversary of the Hungarian uprising – are a reminder that in an age of modern communications, dissent is much harder to contain. At the very least, concern with such social and political issues has a broader base than at any time in the past decade and has already had some success in shaping the official agenda for debate in Eastern Europe.

Gradually, very gradually, Eastern Europe is reasserting its Europeanness, a process that was encouraged initially by the increased contacts between East and West in the 1970s. There are still few illusions about what such a process can achieve, particularly among those in Eastern Europe who are actively working to further it. However, the more Gorbachev tries to appeal to European sentiment in support of his foreign policy, the more he will encourage both the regimes and the peoples of Eastern Europe to search for their reference point in their European heritage, rather than in their recent history.

Gorbachev did not inherit control of the Soviet empire in Eastern Europe in order to see that control waste away, any more than his predecessors had done. Yet, like his predecessors, he faces the problem of how best to manage an increasingly complex and potentially fractious alliance. He has yet to give any clear indication of how he will go about solving it, but his advisers are probably keeping the example of Khrushchev in the 1950s in the forefront of their minds. Hoping to set the Soviet–East European relationship on a more flexible footing, and therefore to maintain Soviet leadership of the alliance more easily, Khrushchev succeeded only in provoking the sort of upheaval he had sought to avoid. With a further thirty years of accommodation with Moscow behind it, Eastern Europe is today a more sober-minded place; all the same, Gorbachev talks persistently of the need for change, and his trips over the past year can only have convinced him that the interests of future stability in the Warsaw Pact require change in several East European countries, not just in the Soviet Union. In considering whether to loosen controls over Eastern Europe or maintain Soviet domination, Gorbachev has little choice but to try to do both.

The Middle East and the Gulf

ISRAEL AND ITS NEIGHBOURS

A key element of strategic developments in the Israeli–Arab and intra-Arab spheres was the complete failure of the Jordan–PLO partnership in 1986. Although it precluded the possibility of any tangible progress in the Palestinian peace process during the rest of the year, this collapse opened the way for other developments – in Lebanon, and in Israel's relations with Jordan and Egypt – to run their course relatively unhindered. The brutal war between Iran and Iraq also continued, unhindered by any sign that Iran was contemplating negotiations as a more sensible way to end the conflict than the far more costly, and less certain, effort to achieve military conquest.

Jordan and the PLO

On 19 February King Hussein brought his second attempted Jordanian–PLO rapprochement since 1982 to an official end in a long speech that provided an unusually detailed review of his approach to the Palestinian question. He emphasized his perception that Jordan could and must act to prevent Israel from expelling Palestinians from the West Bank. Yet, he noted, PLO policy towards Israel had brought about a situation where 'defence of the PLO gradually replaced the basic issue, which was the liberation of occupied Palestinian territories'. After analysing the need for action and the PLO's failure to meet it, Hussein then traced his own efforts to generate understanding between the United States and the PLO, based on his 11 February 1985 accord with Arafat. These, he revealed, had actually persuaded Washington to agree that the PLO should be invited to attend an international peace conference, on condition that it recognized UN Resolutions 242 and 338, agreed to negotiate peace with Israel, and renounced terrorism.

But here Arafat had torpedoed the entire process: 'our brothers in the Palestinian leadership surprised us by refusing to accept Security Council Resolution 242 within this context'. Specifically, the US (and, indeed, Hussein himself) rejected Arafat's demand for recognition in advance of the Palestinian right to self-determination. Consequently, the king concluded, 'I and the government of the Hashemite Kingdom of Jordan hereby announce that we are unable to continue to co-ordinate politically with the PLO leadership until such time as their word becomes their bond'.

Internal PLO schisms and increasing terrorism inside Israel and the occupied territories (often, apparently, the work of Amman-based PLO operatives, and hence likely to embroil Jordan in Israeli retaliatory measures) undoubtedly fortified Hussein's will to break with the PLO. So, too, did the prospect of collaborating for the remainder of the year with the Peres government in Israel, which had proved that it was prepared at least to slow down the Jewish settlement drive.

In ending his effort at collaboration with Arafat, Hussein did not totally renounce a role in Palestinian politics. To the occupied territories he pledged 'we will continue to support you'. In particular, his concern over the dangers of Israeli-enforced (or encouraged) Palestinian emigration from Gaza and the West Bank to the East bank – Jordan – implied that the arrival of such emigrés would weigh heavily upon his kingdom. Throughout the remainder of 1986 he directed his efforts towards generating improved economic conditions and enhancing Hashemite influence in the occupied territories. This effort made a marked contrast with his reaction to the failure of his previous effort to solicit a PLO partnership – in April 1983, in the aftermath of the PLO's removal from Lebanon. Then, Jordan had abdicated any further role. Now, genuine concern to prevent damage to his own kingdom, together with his apparent perception of an opportunity to fill a vacuum and make some progress towards a settlement under his own steam, encouraged Hussein to begin to undertake what the Israelis termed 'Jordanization' of the occupied territories.

The Jordanian move thus signalled the end of one complex dynamic and the start of another. The abortive attempt by Jordan, the PLO, Israel and the United States to settle two issues – Palestinian representation in the peace process, and the nature of an international conference demanded by the Arab side to legitimize and secure broad support for its outcome – now gave way to an even subtler combination of quiet Jordanian–Israeli collaboration in the occupied territories, with the United States' blessing. This was accompanied by a number of complementary processes. One was a parallel Jordanian–Syrian *rapprochement de convenance* – even though Hussein's mid-year attempt to mediate between Syria and Iraq (and so detach the former from its alliance with Iran) was coldly exploited by Syrian President Assad to manoeuvre Tehran into augmenting its aid package to Damascus. Another was a moderately fruitful effort by Israel and Egypt to exploit the lack of steam in the Palestinian issue to repair their own relationship. Lastly, by the end of the year Arafat had successfully launched an effort to make up for his loss of a Jordanian base by reasserting himself on the Lebanese scene.

The 'Jordanization' of the Territories

Neither the return of Hussein's attention to the West Bank and even the Gaza Strip, nor the demise (or perhaps ritualization) of the international conference idea were clearly foreseeable early in the year.

Israel, Egypt and Jordan continued to discuss the conference idea well into 1987 – although Yitzhak Shamir (who had succeeded Shimon Peres as Israeli prime minister under the National Unity Government's rotation agreement) and US Administration spokesmen paid at best reluctant and ambiguous lip service to it. And Hussein's new approach to the occupied territories only began to take clear shape in mid-summer, when a $1.5-billion five-year development plan (which he hoped would be funded in conjunction with moderate Arab countries and the US) was published. This aimed at 'preventing emigration' through enhanced economic development of the Arab sector in the territories and 'strengthening national ties' between the two banks of the River Jordan.

By this time Hussein had closed down all 25 *Fatah* offices in Amman, which had been allowed to open after the November 1984 Palestinian National Council meeting that had heralded the Jordan–PLO thaw. He also expelled Abu Jihad, Arafat's deputy for operations, and even set up his own proxy *Fatah* organization under Abu al-Zaim. He now moved to a series of positive steps inside the territories. Despite the fate of the Jordanian-approved Mayor of Nablus, Zafr al-Masri (murdered in March, apparently by pro-Syrian Palestinians), he acquiesced in the Israeli appointment of Arab mayors for Hebron, Ramallah and Al Bira, the last three West Bank towns still under Israeli military mayors. Jordanian passports were offered to stateless residents of the Gaza Strip, and a Gaza affairs bureau was established in Amman. Jordanian-salaried personnel were installed in West Bank universities and chambers of commerce, the Cairo-Amman Bank reopened (the first Arab bank in the occupied territories since 1967), and a pro-Jordanian newspaper was funded in East Jerusalem.

The process was not without its difficulties. For one thing, the Palestinian population of the territories either remained largely indifferent or – worse from Hussein's and Israel's point of view – continued to express support for the PLO, particularly in student riots in mid-December. The Jordanian initiative had the effect of causing PLO factions in the occupied territories to close ranks around a more radical, somewhat Islamicized policy line. Furthermore, neither the treasuries of the oil-rich Arab states nor that of the United States were fat enough to contribute the money Hussein needed to finance his scheme, and the Gulf states feared angering the PLO. By the end of 1986 Hussein had collected only some $50 million, and lack of funding threatened to slow his entire development plan. And Egypt, which remained committed to PLO participation in any solution, was cool towards the plan, especially in view of Hussein's new and possessive attitude towards the Arabs of the Gaza Strip, whom it had traditionally considered to belong to an Egyptian sphere of influence.

Israel was more co-operative, to the extent of heavily repressing West Bank activists who did not find favour with the king. As for funding, Peres suggested a 'Middle East Marshall Plan' within the framework of the peace process, which called for the industrialized West and Japan to provide financial aid for those Arab states worst hit by the drop in oil-related income – a programme not accidentally tailored to the needs of Jordan and Egypt. Peres' continued boosting of the international conference idea was also evidently co-ordinated with Amman and provided a convenient international alibi: in effect, both sides could claim that any real progress towards a settlement would, of course, be referred to such a framework. What they really intended was explained by Peres in an October interview:

> Let's be honest. The international conference is not needed either by Israel or the US. It is required to bring Jordan to the conference table. . . . The problem is, how to create a niche for Jordan to enter the process. The formula is: make enough of an international conference so Jordan can enter negotiations, and stop it at the right moment, so that negotiations become direct and bilateral.

As the year came to an end, however, recognition grew in Israel that PLO success in re-establishing a political stronghold in southern Lebanon could so inflame attitudes in the West Bank that it might torpedo the entire Jordanization process. It was liable to undo Hussein's efforts to alter the balance of forces between his supporters and those of the PLO in the Israeli-administered territories.

Israeli Foreign-policy Achievements

What Peres generally played down was the immense benefit which the moderate and forthcoming image projected by the international conference and Marshall Plan ideas brought for a series of Israeli foreign-policy initiatives only tangentially related to the Palestinian question. For example, two black African states, Cameroon and the Ivory Coast, renewed diplomatic ties in the course of 1986. Links with the USSR and the Eastern bloc also improved, as a direct result of Israel's display of willingness to involve the Soviet Union in the peace process; after the ice had been broken at an August consular affairs meeting in Helsinki, Peres had talks with Soviet Foreign Minister Shevardnadze at the UN in September, and Poland and Israel opened interest offices in each other's countries in October. There were indications of an opening to China, too – Beijing, after all, would also be invited to play a role at an international conference for Middle East peace. Spain's agreement to establish diplomatic relations in January can also be attributed to Israel's improved international image.

Closer to home, Turkey raised the level of its diplomatic representation in Tel Aviv in June, while Peres' public meeting with King Hassan of Morocco in July (the first between Israeli and Arab leaders since 1981) was noteworthy for arousing so little protest from other Arab quarters – Libya's Col. Gaddafi fumed, but only Syria broke relations with Morocco over it. Predictably, Peres and Hassan accomplished nothing of an overt nature to advance the peace process, but the momentum generated was sufficient to carry Israel and Egypt through the final niggling negotiations over the Taba controversy to the signing of an arbitration compromise on 10 September and a Peres–Mubarak summit in Alexandria the next day. Here too, more lip service was paid to the international conference idea. More substantively, Israel gained a renewed Egyptian commitment to 'normalization' (the first step being the elevation of Cairo's *chargé d'affaires* in Tel Aviv to ambassador status), while Mubarak shored up his image in the United States, ahead of the heavy financial requests he was preparing, without unduly compromising himself in Arab eyes.

Israel's most impressive success in external affairs, however, was not even indirectly linked to the vicissitudes of a Palestinian solution. This was the growing Israeli–American strategic alliance. During the course of 1986 more and more of the new components of this relationship were revealed: joint naval and other military measures; the provision of SDI research funds; congressional approval of funds for conventional-weapons research and development; and initial steps towards American military pre-positioning in Israel. These aspects of collaboration took their place alongside extraordinary congressional efforts to maintain annual aid to Israel at the level of $3 billion for FY 1987, despite severe cuts in the US foreign-aid budget, and the final delivery of $1.5 billion in special economic aid for the Israeli recovery package. Even the most controversial issues, such as American financing for Israel's plans to build the *Lavi* ground-attack aircraft and a new generation of missile boats and submarines, were handled with a degree of collaboration unprecedented in the history of American relations with the Middle East. Indeed, it was the policy flops and scandals that highlighted most strongly the stability of the US–Israeli relationship. Continued reverberations of the November 1985 arrest of US naval intelligence analyst Jonathan Pollard on charges of spying for Israel were not allowed to hinder business as usual. And the Iran arms scandal, however misconceived its strategic rationale, showed how far the two states were co-ordinating policy and action in the Middle East.

Israeli Domestic Improvements

Besides its exploits in strategic and foreign policy, the Peres government ended its tenure of office under Israel's unique experiment in power rotation with a considerable degree of success in areas nearer to home. Inflation was down to about 20% a year in 1986 (it had been nearly 500% when Peres took office in 1984), while most of the dangers of a radical economic slowdown, such as unemployment, appeared to have been avoided. The atmosphere of racial and ethnic rancour that had characterized the final years of *Likud* government also seemed to have been alleviated. Some of these achievements should be attributed to Peres' own forceful personality and dogged determination, but most derived from the very existence of the National Unity Government, which, despite seemingly endless minor crises, survived through to rotation in mid-October.

Indeed, any government other than this full-scale coalition could hardly have weathered the exhausting series of genuine national scandals and crises that unfolded in 1986 after festering under the surface for years. For the first time in Israel's history emigration exceeded immigration; a dangerous water shortage, caused by years of overpumping, presaged alarming problems for the future of agriculture; many of the country's banks were revealed to be corruptly managed; and a series of security scandals involving the vaunted General Security Service, the aftermath of the Pollard affair, and former nuclear industry worker Mordechai Vanunu's revelations about Israel's nuclear potential, as well as some aspects of the Iran arms scandal, all encouraged serious doubts about government control over security and intelligence operations.

All in all, however, the first two years of the rotation experiment had to be considered a success for both major parties, and particularly for Peres. This did not go unnoticed by Shamir, who was careful during his first months in office to take no action – e.g., increasing Jewish settlement in the West Bank – that might be interpreted at home, or in Washington, Cairo or Amman, as undermining the goodwill Peres had built up. Indeed, in anticipation of a February 1987 trip to Washington, Shamir publicly hinted at his own readiness to pursue 'Jordanization'. Certainly, as long as Peres' popularity remained high, and he had troubles within his own party, Shamir could hardly risk any initiative that might unravel the coalition government structure.

By early 1987, then, Israel appeared to have weathered the worst of the internal crises – economic, social and, ultimately, electoral – that had come to a head in 1984. In Arab eyes it certainly appeared unified and strong (Vanunu's revelations about its nuclear arsenal only strengthening the country's deterrent image), with a solid American ally to back it, and under virtually no pressure from any side to make far-reaching concessions to further the peace process. The two remaining years of the Reagan Administration in Washington seemed to offer only more of the same for Israel. Israel's cultivation of Iran was viewed with disquiet by many in the Arab world. The strengthening of a revolutionary state of this kind not only put in jeopardy the emerging, pragmatic peace process in the region but also seemed to threaten the political regimes of a number of Arab states. Controversy over these questions was sufficient to maintain the disunity of the Arab world itself.

Syria and Lebanon

One area that appeared by late 1986 to be particularly susceptible to unexpected developments was the Israel–Syria and Israel–Lebanon borders. An atmosphere of tension pervaded the Israeli–Syrian front throughout the year, nourished by Syria's repeated declarations that it was approaching 'strategic parity' with Israel and its continued support for international terrorist acts. Israeli apprehensions were also fed by the fear that Syria's serious economic difficulties might tempt the politically isolated Assad to launch an attack in order to rally inter-Arab support for his regime. (At several points during the year Syrian foreign-currency reserves were down to a few weeks' exports, and rationing of electricity and essential goods occurred frequently.) On the other hand, it seemed only a matter of time before economic constraints forced Damascus to reduce its armed forces, which had grown by two divisions since the 1982 battles with Israel in Lebanon. Lastly, the Assad regime had troubles with internal terrorism, most of which was apparently Iraqi-sponsored, and which ostensibly should restrain any desire to undertake new strategic initiatives.

Thus the ailing Assad, beset by succession squabbles, was perceived by Israel as having little to lose from launching a limited strike aimed at capturing part of the Golan Heights, whilst raining his new, longer-range surface-to-surface missiles on the north of Israel. Israel, with its own economic difficulties and its unwillingness to face further manpower losses after its fighting in Lebanon, would be reluctant to launch a heavy counter-strike into the extremely dense Syrian fortifications between the Golan Heights and Damascus. For its part, Syria feared that, while it was still alone in confronting Israel on the defunct eastern front, the latter might be tempted to launch a preemptive strike to destroy its military potential. It also saw the hand of Israel in almost every new development in Lebanon, an area of influence it claimed for its own. Its anger over Israel's interception, on 4 February, of a Libyan executive jet carrying Syrian officials to Damascus – mistakenly thought to have Palestinian terrorist leaders on board – explained at least a little of its readiness to sponsor anti-Israeli terrorist acts in the ensuing months, most dramatically the 17 April attempt to blow up an El Al aircraft departing from London.

But as 1986 came to a close it was Lebanon that appeared to harbour the greatest immediate dangers of instability. Not only were Syria and Israel involved, but they both had to contend with the Maronites, the militant Shi'ites and, less predictably, the Arafat-led PLO and Iran.

Essentially, 1986 saw an upsurge in extremist Shi'ite and Palestinian presence and influence in Lebanon, at the expense of nearly all other interested parties except one relatively new arrival, Iran. Syria tried to assert its hegemonic aspirations early in the year and failed. The tripartite agreement it sponsored among the Shi'ite *Amal*, the Maronites and the Druze under Walid Jumblatt, signed in Damascus on 28 December 1985, had been designed to effect constitutional reform at the Maronites' expense and to enhance Syria's influence. Within two weeks it had collapsed as a result of Maronite opposition (President Gemayel himself travelled to Damascus to reject it), and Syria's man in the Maronite Lebanese Forces, Elie

Hobeika, was deposed in favour of the more moderate Samir Geagea. In the ensuing months Syria sponsored a series of terrorist attacks and coup attempts against the Maronite leadership, as well as making various military moves which Assad did not bother to co-ordinate with President Gemayel – a battalion was sent to Mash'ara in the south on 6 June, and on 6 July a small Syrian troop contingent entered troubled West Beirut. Meanwhile it continued to support the relatively moderate Shi'ite *Amal*, but soon found itself confronting a complex web of alliances over which it had little control. To redress the situation, Syria was forced at the end of February 1987 to send over 7,000 troops into West Beirut to halt the intense fighting that had broken out between the Iranian-backed *Hizbollah* and other militias, particularly *Amal*.

Part of the difficulty it faced was caused by Iran's attempts to assert its influence in Lebanon – without co-ordinating policy with Damascus – through the extremist *Hizbollah*. This group – which had a relatively small militia of 4,000 fighters, less than a quarter of them deployed in the south – declared its provocative goal of turning Lebanon into an Islamic fundamentalist state and liberating Jerusalem via Lebanon. It thereby set itself against *Amal*, which limited its objectives to enhancing Shi'ite influence within the Lebanese ethnic mosaic and expelling the Israeli proxy presence from the southern security strip. Iran also evidently lent its support to several shadowy Shi'ite terrorist groups like *al-Jihad al-Islami*, whose continued abductions of western hostages (including the British negotiator Terry Waite) drew the spotlight of international publicity upon Lebanon and emptied it of Americans and Europeans by February 1987. Just two months earlier, in December 1986, it had become clear that a new external actor had arrived on the scene when Iranian diplomats in Lebanon and Syria negotiated a cease-fire between *Amal* and PLO contingents in the southern village of Maghdousheh, to be policed by *Hizbollah* militiamen.

The *Amal*–PLO fighting also indicated that an armed Palestinian presence in Lebanon had been renewed. By mid-1986 the mainstream PLO under Arafat had been ejected from Jordan and was under pressure from the government of Tunisia – unhappy since Israel had bombed PLO headquarters there in October 1985 – to thin its forces. Needing a territorial base with a sizeable Palestinian population near Israel in order to maintain both an option for armed struggle and his own leadership status, and perceiving a new power vacuum in Lebanon, Arafat acted to reintroduce his *Fatah* legions to West Beirut and the south. Here, in fairly typical topsy-turvy Lebanese fashion, he was aided by the Syrian-supported rebels of Abu Musa, who evidently preferred an Arafat-PLO presence to none, and by the Maronites, who saw the return of their old enemies, the Palestinians, as a convenient trump card to play against their newer rivals, *Amal*.

This set the stage for the camps war of 1986. The Sabra camp having been decimated by *Amal* in the 1985 camps war, some 9,000 lightly-armed Palestinians, mostly mainline *Fatah*/PLO, defended Shatila and Bourj al-Barajneh in West Beirut and Rashidiyeh camp near Sidon, against the better armed and more numerous *Amal*, while Syria looked on in embarrassment and Iran exploited the opportunity to make new friends. The result was a stalemate which in the circumstances, and despite more than 2,500 casualties, constituted a victory for the PLO and a setback for *Amal*, to whom any renewed PLO presence in Shi'ite-dominated territory represented a threat. Almost incidentally, Arafat's now total preoccupation with Lebanon was facilitating the Jordanization process in the West Bank and Gaza Strip.

Amal could be expected to regroup for another attempt to prevent a Palestinian armed return, particularly to the south, where the PLO presence inevitably evoked memories of Israeli invasions. As for Israel, it found the aggressive tactics of *Hizbollah* a worrying enough challenge to the southern security strip, without having to consider the PLO. In September and December several large-scale *Hizbollah* attacks on outposts manned by Gen. Antoine Lahad's South Lebanon Army (the Israeli proxy force) took a heavy toll in SLA lives and morale. The *Hizbollahi* were evidently supported by Shi'ite contingents from the Lebanese Army and trained and armed by Iranian *Pasdaran* stationed in the Beka'a valley. With significant support from the Israeli forces Lahad was able to regroup, but Israel was forced out of its late 1985 and early 1986 complacency about what had seemed to be the unqualified success of the southern security strip in preventing terrorist incursions and *Katyusha* rocket bombardments into northern Israel. It now began to evince genuine concern that the combination of militant Shi'ites with Iranian support, and perhaps PLO forces too, might eventually require it once again to play a more active military role in Lebanon. Most Israelis, recalling 1982 and its aftermath, recoiled at this prospect. One interim measure adopted in late 1986 was to provide unsolicited aid to *Amal* (bombing and strafing PLO and *Hizbollah* positions) in its struggle to retain a measure of supremacy in the south. But *Amal*'s own implicit admission that it was losing ground only reinforced the Israeli perception that the southern security strip – once criticized by sceptics as a poorly-conceived substitute for allowing *Amal* to police the Lebanon–Israel border – was, for the time being, the best solution for a bad situation.

Troubled Times in Egypt

In November and December President Mubarak marked five years in power with a series of actions that reflected something of the domestic dangers threatening Egypt's future. On 9 November he installed a new government – headed by an economist, Dr Atif Muhammad Nagib Sidqi – in the hope of finding new answers to Egypt's seemingly insurmountable socio-economic problems: a population growing by a million every eight months (Cairo's is expected to double to 20 million by the year 2000) while water and arable land resources were nearly static; an external debt exceeding $38 billion, with annual payments of principal and interest reaching $4.3 billion (up from $800 million in 1982); and a rising tide of Islamic fundamentalist and leftist agitation. Mubarak had already had to use troops in February to put down a gendarmerie revolt in Cairo sparked off largely by economic neglect. In early December the government arrested 33 members of *al-Jihad* (including four junior army officers) and followed this up in mid-December by seizing 44 Marxist extremists from the 'Revolutionary Current' group centred in Giza. Late in the year, too, Mubarak embarked on a concerted effort to improve Egypt's debtor status with the West; in December he visited France and West Germany in the hope of garnering new loans on favourable terms and recruiting pressure on the United States and the IMF to ease his country's interest payments. But he postponed a trip to the US scheduled for the second half of 1986, largely in recognition of the difficulties he knew he would face in the aid sphere. Here the Iran arms scandal, whatever its genuine strategic impact on Egyptian perceptions of the US role in the region, provided an opportunity for Mubarak (and King Hussein) to bewail Washington's loss of credibility, perhaps in the hope of improving Cairo's chances of receiving compensation in the form of enhanced economic aid.

Having cancelled the January 1986 *Bright Star* joint exercises with American forces after the *Achille Lauro* affair (when the US unilaterally intercepted an Egyptian plane carrying the terrorist Abu Abbas), Egypt reactivated its strategic collaboration with the United States with the *Sea Wind* joint air and naval manoeuvres in late August 1986. The fact that these took place close to the Libyan coast enhanced the overall impression that ultimately Mubarak would still be willing to go a long way to accommodate US interests in the region, whatever his misgivings over US aid and US dealings with Iran.

Mubarak himself, in a fifth anniversary interview, called 1985 his 'hardest year in office' because of the *Achille Lauro* affair, and complained of the heavy security restrictions under which he lives ('I feel like I'm in the Bastille') – hardly a reflection of the sort of ebullient, innovative policies with which his predecessor, Anwar Sadat, had so improved Egypt's strategic standing in the region. But Mubarak's low-key approach did restore Egypt to a position of some influence on the inter-Arab scene (in January 1987 he participated, rather triumphantly, in an Islamic Conference summit for the first time in eight years), and he could still rely on the military to support him against domestic upheavals. Looking further ahead, though, the cold demographic and economic facts of Egyptian life remained largely neglected by a regime too frightened of arousing mass opposition to take the painful steps that might prevent the country's political system from collapsing at some time in the not-too-distant future.

A Shift of Emphasis

While the question of Israeli–Arab opposition remains as a background motif in Middle East affairs, the central focus has switched to inter-Arab conflict. Not only does the Iran–Iraq war of attrition grind on with little sign that it will reach a climax soon, but the continuing battle in Lebanon has once again pitted Syrian troops against their co-religionists. And the attempt by the Arafat wing of the PLO to re-establish itself in Lebanon was being opposed by both *Amal* and *Hizbollah.*

If Israel was still unable to find a path to peace with its Arab neighbours (except for Egypt), there was little sign that any of them would contemplate going to war with Israel in the near future, either to support Palestinian claims or to recover territory claimed by one state or the other. Nor was there much coherence in Arab affairs: Syria might support Iran in the Gulf war because of its opposition to Iraq, but it also opposed Iranian-supported extremists in the Lebanon. Jordan might make common cause with Syria to prevent the PLO from reasserting itself in Lebanon, but it continued to oppose Syrian-backed Iran in the Gulf war. And all the states in the area were preoccupied with internal difficulties. It was difficult to envisage much significant change, for better or for worse, in the Middle East in the coming year.

A CRITICAL YEAR IN THE GULF

The sixth anniversary of the outbreak of the Gulf war passed without the launching of the much feared 'final offensive' which elements of the Iranian leadership had been promising all through the summer of 1986. The respite proved to be short-lived, though. By the end of the year Iran appeared to have overcome – at least for the time being – the damage caused by reduced oil revenues and political infighting, and the early weeks of 1987 provided another jolt for those inclined to place their faith in Iranian war-weariness and the prospect of an inconclusive stalemate.

Despite the questions that have arisen about Iraqi morale and the stability of the Baghdad regime, and those raised by Iran's domestic problems, neither belligerent seems ready to give up the struggle. The readiness of the Iranian leadership to deal with the US to secure arms was no indication of a new 'moderation' in Tehran and clearly implied no diminution in its determination to punish Saddam Hussein, who continues to dominate the Iraqi scene.

Iran's strategic importance and its influence elsewhere in the Middle East led a number of states to seek closer relations with Tehran. Nevertheless, France remained outspoken in its commitment to Iraq's survival, as also did the Soviet Union. The US was

TURKEY
USSR
Caspian Sea
Sulaymaniyah
Tehran
Baghdad
Mehran
IRAN
Tigris
Isfahan
Euphrates
IRAQ
Khorramshahr
Basra
Shatt al-Arab
Abadan
KUWAIT
NEUTRAL ZONE
Faw Peninsula
Ganaveh
Kharg Island
The Gulf
SAUDI ARABIA
BAHRAIN
Larak Island
Riyadh
Strait of Hormuz
Sirri Island
Ras al-Khaymah
QATAR
Sharjah
Dubai
Abu Dhabi
0 100 200 miles
0 100 200 300 kms
UAE
OMAN

revealed to have been supplying arms to Iran, though it had long claimed not to be involved in the conflict, and some observers drew the conclusion that the US had actively decided to prop up Iran's cause. This was not US policy, but the affair served to illustrate the complex diplomatic ramifications of a war which continues almost in defiance of the wishes of the international community.

Military Developments

In mid-1986 Iran's threats caused considerable apprehension, both within the region and further afield. The attention devoted to them in the international media was not simply the usual coverage that the anniversary of the outbreak of the Gulf war regularly attracts, nor just the result of hasty analysis: there were good reasons to think that the war was entering a more ominous phase. The latest threats appeared to warrant more serious consideration than previous ones because of an apparent deterioration of Iraqi morale and because of Iran's worsening economic position: a combination which seemed to give Iran a double incentive to try for a rapid end to the struggle.

In February 1986 Iranian forces had succeeded in crossing the Shatt al-Arab (Operation '*Dawn* 8') and establishing a foothold on the Faw peninsula, whence heavy Iraqi counter-attacks failed to dislodge them. Later the same month, an offensive in northern Iraq ('*Dawn* 9') managed to capture small amounts of territory near Sulaymaniyah. In mid-May, in an attempt to offset their psychological defeat at Faw, Iraqi forces seized Mehran, in the central sector of the front; but Iran spurned the offer of an exchange for Faw and, some weeks later, retook Mehran with considerable ease (Operation '*Karbala* 1'). In the West – and almost certainly in Iran too – it was widely believed that these setbacks had done serious damage to Iraqi morale. And Saddam Hussein's 'open letter' to the Iranian leadership, urging them to show restraint and abandon their plans for a 'final offensive', could only strengthen such an impression.

Iran had another reason for using this opportunity to launch a 'final offensive' to try and win the war. International oil prices had fallen; so, too, had its own oil exports – as a result of successful Iraqi air attacks on Kharg Island, on mainland pumping stations and, on 12 August, on Sirri Island at the mouth of the Gulf. Iran's income was therefore certain to fall far short of the figure of $18.6 billion projected for the financial year beginning in March 1986, and estimates of the likely actual figure were falling daily. No doubt there were therefore many in Tehran who were urging that this was the moment to take decisive action, rather than let Iran lose the initiative and spend itself slowly into impotence.

On the other hand, it could not be ignored (and, it seems, was not ignored) that Iran still lacked the weapons and logistic capacity to mount a sustained offensive. Its forces might therefore have to resort once more to the discredited tactics of human-wave assaults in order to breach Iraq's strong defensive system. If a major offensive was seriously contemplated, therefore, it seems that prudent counsels pointed out the political risks inherent in staking everything on an operation which would probably have failed.

In the event, there was no offensive, despite signs of a build-up of manpower. Instead Iran mounted small operations in Kurdistan and at the mouth of the Shatt al-Arab in September, and staged another limited offensive in northern Iraq in November. Unable to reply on the ground, and no doubt aware of Iran's economic problems, Iraq embarked on a bombing campaign which was more successful than its previous efforts. The success at Sirri was followed up with heavy attacks on Kharg in October, and on 25 November the Iraqi Air Force showed that its reach extended to Iran's newest terminal at Larak Island in the mouth of the Gulf. Raids were also mounted on industrial and infrastructure targets in Iran. In retaliation, Iran launched missiles against Baghdad, which confirmed the trend towards intensified reciprocal attacks on population centres.

Arms for Iran

Iran's efforts to overcome its materiel disadvantages were meanwhile achieving some success, albeit at the cost of further diversification of its equipment inventory. Despite its official denials, China was known to have supplied large quantities of heavy equipment, including some 50 J-6 fighters. New, US-made weapons were identified in the Iranian armoury, and there were reports of substantial Israeli shipments reaching Iran.

In November this rearmament drive provided the context for the year's most publicized development – the revelation that for some eighteen months the United States had been supplying arms worth an estimated $12 million to Iran. What shocked US domestic and international opinion was not just the secrecy and questionable legality of this, but also the fact that it undermined the credibility of loudly declared US policies: of non-involvement in the Gulf war; of not supporting states linked with international terrorism; and of refusing to deal with hostage-takers. Regional friends of the United States were further embarrassed by the evidence of close Israeli involvement in the operation.

The declared rationale for the supplies was the need to encourage those 'moderate' elements in Iran who might ultimately come to favour closer relations with the United States. But this story was not altogether persuasive. It soon emerged that the arms shipments had become linked to an effort to persuade Iran to use its goodwill to secure the release of American hostages held in Lebanon, and that a potentially embarrassing arms-for-hostages deal had been presented as a coherent strategic plan. Doubts were also raised as to whether, if the (simplistically-termed) 'moderates' were to come out on top in Tehran, they would really bear US regional interests in mind when formulating their policies.

To be sure, the volume of arms supplied was small; nevertheless the deliveries raised questions about the wisdom of being seen to assist the Iranian war effort at such a delicate juncture. Even if the US missiles did no more than improve certain aspects of Iran's capability (such as its air defences), they were likely to have the damaging side-effect of legitimizing arms deliveries from other suppliers. Israel, which had acted as a conduit for the US shipments, was also selling much larger quantities of basic hardware to Iran on its own account; China, too, had become an important source of weaponry; and the Soviet Union was also reported to be involved. Any US criticism of others who were easing Iran's logistic problems could not now have anything but a very hollow ring. True, there were signs that the US was still committed to preserving the balance in the northern Gulf as far as possible; an investigation of the Iranian deal revealed, for instance, that the CIA had for some time been supplying satellite intelligence to Iraq. All the same, this news could not dispel the air of confusion that now hung over Washington's Gulf policy.

At the beginning of 1987, observers were given good cause to speculate on the effects of Iran's arms purchases on its military performance. No sooner had it been confirmed that Iran's '*Karbala* 4' offensive near Khorramshahr in late December had failed, than fighting broke out again in the same sector of the front. After some initial confusion, it became clear that the forces involved in '*Karbala* 5', (launched on 9 January) had, at the cost of some thousands of casualties, managed to cross the Shatt al-Arab and seize a small area of Iraqi territory close to Basra. Iraq managed, however, to contain another offensive ('*Karbala* 6'), which was mounted within days in the central sector of the front near Sumar. As the Revolutionary Guards consolidated their positions near Basra, Iraq responded by dramatically stepping up its air attacks on civilian targets, a strategy which was unlikely to have much effect, and Iraqi counter-attacks in the southern sector did succeed in eroding some of Iran's gains. The thrust towards Basra may have been checked and the attackers kept well clear of the main defensive lines near the city, but Iranian forces had once again shown that they could catch Iraqi defenders unawares and inflict on them an unacceptably high (and potentially demoralizing) level of casualties.

The Situation in Iran

Iranian offensives in the opening days of the year fit into an established pattern that is dependent on the weather. Military considerations apart, however, the January 1987 operation may also have had other objectives as well: to impress the delegates to the summit of the Islamic Conference Organization (which opened in Kuwait on 26 January), and to show the Iranian population that the recent political turmoil in Tehran had not undermined the regime's determination to fight the war to a successful conclusion.

In June 1986 the Speaker of the *Majlis* (parliament), Rafsanjani, made the significant remark that there were 'two powerful factions' in Iranian politics, which in effect constituted separate political parties and held conflicting views on a wide variety of issues. Because of its very generality, this remark must be taken as an oversimplification, but events some months later nevertheless clearly demonstrated the basic accuracy of his observation. The arrest in October of Mehdi Hashemi, head of the organization dedicated to exporting the Islamic revolution, provided clear evidence of factional differences over the principles that should underlie Iran's relations with the world. The guiding hand behind the arrest of Hashemi, and others similarly charged with subversive activities, appears to have been Rafsanjani's. This, plus Hashemi's close links with Khomeini's designated successor, Ayatollah Montazeri, prompted rumours that the latter's political status was being deliberately undermined (although Rafsanjani was quick to dismiss speculation about the affair's effect on relations between Khomeini and his spiritual heir).

The revelations of covert US–Iranian dealings were a by-product of this affair: the story of Robert McFarlane's visit to Tehran became public because individuals opposed to any relaxation of revolutionary principles leaked details to a Lebanese newspaper. This news had immediate repercussions in Tehran. The fact that Rafsanjani, himself deeply implicated in the deal, initially sought to deny any official involvement in the venture was indicative of the level of public concern about, and opprobrium attaching to, contacts with the United States. Critical voices immediately clamoured for a detailed investigation of the affair.

It would appear that Ayatollah Khomeini's intervention was instrumental in preventing a more serious political crisis. Although Khomeini had ordered an inquiry into the activities of Hashemi and his allies, he now enjoined those demanding a similar inquiry into US–Iranian contacts to refrain from causing 'schism'. At the same time he emphasized that the deal did not and could not signify a fundamental shift in Iranian attitudes towards the United States; others, including Rafsanjani, were quick to echo his remarks. Officially, the arms deal was portrayed as a moral and tactical victory which had gained vital supplies for Iran and demonstrated that Washington recognized the Islamic revolution's importance. Rafsanjani was able to take the credit for some astute diplomacy, but he was nevertheless virtually compelled to disavow any further intimacy with Washington. Hashemi, on the other hand, was thoroughly discredited by his televised public confession. Moreover, his statement that he had 'abused the confidence' of Ayatollah Montazeri probably reflected an official attempt to repair some of the damage that had been done to the latter's standing. Montazeri's office quickly denied any association with Hashemi's group. Even so, notwithstanding the publicly declared support of leading officials, and the Friday prayer-leader's praise of his piety and learning on 12 December, it is doubtful whether Montazeri will ever enjoy the political power that Khomeini now enjoys.

It remained uncertain just how much opposition to the war there was in Iran. Certainly many were frustrated by the economic turmoil and the shortages of basic goods, including petrol; their sentiments were openly expressed by former prime minister Mehdi Bazargan, notably in a letter of 27 August calling for an end to the 'ruinous war'. Nevertheless, the hard core of Iran's leadership, and substantial sections of the population favoured a continuation of the war. In September 1986 Khomeini himself inveighed against those who favoured peace or arbitration. Later, in the wake of the disclosures about Iranian co-operation with Washington, he reiterated that there could be no compromise on the goal of defeating the Ba'ath state, which sentiment others duly repeated. As long as Khomeini lives, Iranian acquiescence in a settlement which allows Saddam Hussein to survive is difficult to imagine.

The Situation in Iraq

Yet the Iraqi president would appear to be firmly in command. It is true that the tight security in Iraq makes it well-nigh impossible to gauge accurately the level of public disaffection; nevertheless, foci of discontent are difficult to identify. Saddam Hussein briskly strengthened his position at the extraordinary session of the Ba'ath Regional Command in July 1986, and political control of the Armed Forces continues to be rigid. Indeed, one can say little more than that the casualties and the increasing economic strictures imposed by the war are unpopular, and that many in the Armed Forces must resent Saddam Hussein's readiness to blame them for Iraq's military reverses. But heavy surveillance, the serious external threat and the sheer difficulty of the president's task must greatly reduce the incentive for other aspirants to power.

Baghdad's economic position was complicated in 1986 by the falling price of oil, which made it necessary to impose austerity measures. The situation is likely to remain bleak, despite oil price increases and the prospect of higher oil exports. Nevertheless – and even though Iraq will probably continue to seek deferment of the repayments on its commercial debt – its leadership can be reasonably certain that the strategically-motivated financial support of Arab Gulf states will not suddenly cease. While it is clear that the states of the Gulf Co-operation Council wish to conduct normal relations with Iran, it is equally clear that they consider an Iranian victory to be an undesirable basis for this. Naturally, the options available to individual Gulf regimes have varied. But whether, like Kuwait, they are constrained unambiguously to support Iraq and therefore suffer Iranian vituperation, or whether they can afford the luxury of 'reinsuring' by selling oil to Iran, like Saudi Arabia, they cannot ignore the dangers that an Iraqi collapse would expose them to.

The International Arena

Progress towards better US–Iranian relations may have been frozen for the time being, but the process of rapprochement could in any case never have involved open US support for Iran's war aims.

Equally, the Soviet Union (for whom the desirability of close ties with Iran is axiomatic) and France made it clear that their own dealings with Iran did not imply the abandonment of a long-standing policy of military assistance for Iraq. In September 1986 it was agreed that sales of Iranian natural gas to the Soviet Union would be resumed, and in December the two countries concluded an economic protocol; nevertheless, Soviet official criticisms of Iran's stand on the war in early 1987 were unusually explicit. The French government had made efforts to meet Iran's conditions for improved relations, by expelling the leader of Iran's *Mujaheddin* opposition group and many of his followers in June 1986, and by repaying part of an outstanding $1 billion debt. But this 'normalization', as both Paris and Tehran noted, did not mean a reduction in aid to Iraq.

Nor was Turkey prepared to jeopardize its relations with Iraq, a fact that injected a note of tension into the Turkish–Iranian relationship in 1986. Turkey enjoys close economic ties with both belligerents, and has in many ways profited from its neutrality; but it has also been co-operating openly with Iraq over the mutual problem of Kurdish dissidence, and has indicated its concern for the security of pipeline links between the two countries (a second pipeline is scheduled to come on stream in 1987). Since Turkey seems more or less explicitly to have prevented Iran from exploiting Iraq's traditional problem of Kurdish dissidence more fully, it is not surprising that senior Iranian figures should have accused it of compromising its neutrality in the war. On the other hand, the possible construction of an Iranian–Turkish pipeline was mooted.

There continues to be very little evidence of any international co-operation to bring the war to an end or to limit its effects. The sharp increase in attacks on merchant shipping in the Gulf during 1986 provoked only expressions of concern. Nor was the UN Security Council able to muster the consensus required for a resolution condemning the use of chemical weapons in the Gulf war, despite the fact that a UN report specifically (and for the first time) branded Iraq as a user of such weapons in March 1986. Instead, members of the international community have restricted themselves to the not unimportant sphere of propping up Iraq with arms and financial aid. Worried observers must hope that this will help Iraqi morale to continue to stand up against the efforts of an opponent who remains committed to a military solution, and whose own resources are likely to be boosted by the rise in oil prices. Beyond this, there seems to be very little that they can do.

One must be wary of automatically translating Iran's successes on the ground and the apparently buoyant morale of its forces into a weakening of Saddam Hussein's position. The collapse of the Iraqi army has been predicted all too often without it coming to pass. Nevertheless, Iran's resourcefulness and bellicosity cannot but unnerve many in the region. There is felt to be a real danger that even a partial collapse of the Iraqi front, never mind a change of regime in Baghdad, could serve as a focus for religiously-inspired disaffection as far afield as Egypt. It is not possible to say how long Iran will be able or willing to go on throwing its manpower at Iraq's defences, nor to tell how long those defences will continue to hold. As ever, all predictions are in the short term held hostage by the twin imponderables of Iraqi morale and Iranian logistic capacity.

South and South-west Asia

THE CONTINUING WAR IN AFGHANISTAN

Afghanistan may not yet have become the Soviet Union's Vietnam, but the cumulative costs of intervention there have made it, in the words of Soviet leader Mikhail Gorbachev, a 'bleeding wound'. As a result, the Kremlin has been voicing an increasing willingness to find a political 'settlement' to the problem, although it was only at the end of 1986 that such statements began to be reflected in action. In a flurry of political initiatives, Moscow and its puppet regime, the People's Democratic Party of Afghanistan (PDPA), unfolded what appeared to be a Soviet-designed plan for 'national reconciliation' between the PDPA and the opposition, the *Mujaheddin*. This called for a 'coalition government of national unity' and an agreement, backed by an international guarantee, between the PDPA and Pakistan, which has provided logistic support to the *Mujaheddin* and taken in over three million Afghan refugees.

In late December the new PDPA leader, Dr Najibullah, followed this up with the declaration of a six-month unilateral cease-fire by PDPA forces and, presumably, the 120,000 or so Soviet troops who have sustained the PDPA in power since invading Afghanistan in late December 1979. He asserted the PDPA's 'deep respect' for the religion of Islam and proclaimed a general amnesty for the opposition forces. He called on opponents living abroad, and the *Mujaheddin* leaders, to enter a dialogue with the PDPA in order to formulate a new constitution, elect a national assembly and participate in the government. Further, he promised that an 'attractive' timetable for the withdrawal of Soviet troops, already agreed between his government and Moscow, would be submitted to the February round of UN-sponsored peace talks between the PDPA and Pakistani governments in Geneva.

Certain conditions were attached to this plan, however: a recognition of the 'irreversibility of the April revolution' (that is, the violent 1978 coup which brought the Soviet-backed PDPA to power); a continuation of the PDPA leadership; a strengthening of Afghan–Soviet ties; and an appropriate response from the opposition.

Given that the first two conditions are anathema to the Afghan resistance, and that nine years of bloodshed have made any reconciliation between it and the PDPA unlikely, the *Mujaheddin* promptly and predictably rejected the whole plan, including the cease-fire, as a 'fraud'. They deemed it yet another ploy to enable the Communists to achieve what they had so far been unable to gain by military means, and vowed to fight on for the unconditional, total withdrawal of Soviet troops and the right of Afghans to determine their own future. The *Mujaheddin*'s foreign backers viewed the plan with scepticism, for the same reasons. What is now in question is whether the Kremlin is prepared to sacrifice its puppet regime in order to achieve a viable settlement of the problem and pull out all its troops.

There is no doubt that the Afghan problem is one of the most painful legacies of the Brezhnev era. Soviet casualties are estimated by Western intelligence sources as at least 35,000 since 1979, and towards the end of 1986 Soviet material costs rose sharply to average more than $US 15 million a day. Afghanistan has also undermined the Soviet Union's prestige in the world and remains an important obstacle to the improvement of its relations with many Islamic states, China and the West. And the USSR and its Afghan surrogates have not yet been able either to build a viable PDPA government in Kabul or to extend their shaky hold to other main cities and strategic areas. By the PDPA's own admission, the *Mujaheddin* still control at least 'two-thirds' of the country, with widespread access to the major urban centres, including Kabul.

The Gorbachev leadership seems at last to have concluded that conquering Afghanistan militarily would bring total ruin to the country and involve even greater cost to the Soviet Union than it has had to pay so far. Seeking to make a break with the Brezhnev era, to revitalize Soviet society, and to improve the USSR's image abroad, Gorbachev has good reasons to look for some way out of the Afghan problem as soon as possible.

But the problem is a complex one. In seeking a settlement based on conciliating conflicting interests, Gorbachev apparently believes that a balance can be struck between what the Soviet Union is prepared to accept, and what is acceptable to the Afghan people and the international community. On the one hand, he has retained his predecessors' basic objective: to ensure the long-term survival of the PDPA regime and Soviet influence in Afghan politics (which was why the USSR invaded Afghanistan in the first place). On the other, he has found it necessary to pursue his ends through political more than military means. This effort has taken two forms. One has been to try to split the resistance and expand the power-base of the PDPA by winning over opposition elements who are weary of the war and might be susceptible to PDPA and Soviet promises. Another has been to induce the *Mujaheddin*'s foreign supporters to curtail their aid to the resistance in return for a settlement which provides for Soviet troop withdrawal and a more broadly based regime under PDPA leadership. Moscow's prime target is Pakistan, whose acceptance of such a deal would also influence the attitudes of Iran (which now has 1.5 million Afghan refugees) as well as China and the United States.

The Kremlin has some reason to be optimistic about a 'non-interference' agreement between the PDPA regime and Islamabad. The Geneva talks between the two have concentrated on such issues as an internationally guaranteed agreement, 'non-interference', the repatriation of Afghan refugees, and a timetable for a withdrawal of Soviet troops. Substantial progress has been reported on the first two issues, and a settlement now largely hangs on a resolution of the last. Thus, so long as Pakistan does not make the legitimacy of the PDPA regime the predominant issue, the USSR can be hopeful of obtaining an overall settlement which treats the symptoms of the problem, but not its cause. This would give it a resolution of the Afghan problem largely on its own terms, and a Soviet troop withdrawal might be possible within a shorter period than Gorbachev's predecessors had anticipated. At worst, it would mean Afghanization of the war, with much less direct Soviet military support needed to sustain the PDPA than is currently the case.

This change in the Soviet strategy became apparent towards the end of 1985 and early in 1986. Having set the tone in his report to the 27th Party Congress on 25 February, Gorbachev gave a clear outline of his plans in a widely publicized speech in Vladivostok on 28 July. As evidence of the Kremlin's sincere desire for a settlement and withdrawal of its forces, he promised to pull six Soviet regiments (about 7,000 troops) out of Afghanistan before the end of the year. But he emphasized two important prerequisites for total withdrawal: first, armed hostilities against the PDPA must stop in order to allow it to broaden its social base; second, the PDPA's power base must be built up, not only through military pacification and Party monopoly of power, but also through 'national reconciliation'. This should involve co-opting opposition elements prepared to accept the legitimacy of 'the April national-democratic revolution' (and therefore the PDPA's leadership) and to 'participate sincerely in the nationwide process of the construction of a new Afghanistan'. Only on such conditions, and when 'a political settlement is finally worked out', would the Soviet Union be ready to withdraw its troops. Even then it would be a 'stage-by-stage' process, according to 'timetables' which 'have been agreed upon with the Afghan leadership'.

The Course of the Fighting

In 1986 the Soviet/PDPA forces (mostly Soviet, since the Afghan Army commands no more than 30,000 troops, many of them unreliable) stepped up what may be termed a strategy of forward fighting which emphasized frequent limited offensives, intense firepower and staying power, rather than the flexible, defensive response operations which had largely characterized Soviet tactics in the first six years of fighting. The prime objective was to destroy the *Mujaheddin*'s strongholds, cut off their supply and infiltration routes from Pakistan, and deprive them of social and material support from the population. With Soviet elite units (*Spetsnaz*) spearheading the operations, Soviet/PDPA forces began their most intense offensives of the war without waiting for the usual winter lull in the *Mujaheddin*'s activities to end – an unusual departure from form. In early February, they launched a major attack in Nangarhar province, followed by larger offensives from early March to mid-April in all the eastern and south-eastern provinces bordering Pakistan, from Kunar to Paktia and Kandahar. The result was heavy civilian casualties and an increase in the flow of refugees into Pakistan. Two of the *Mujaheddin*'s important bases at Barikot and Zhawar on the border with Pakistan were destroyed. The Soviet/PDPA forces continued with one such full-scale operation each month, mainly in the provinces bordering Pakistan, using indiscriminate bombing and mine-laying wherever they detected *Mujaheddin* strongholds and suspected popular support for them.

There was also a dramatic increase in PDPA military and subversive operations in Pakistan's North West Frontier Province (NWFP) and Baluchistan, where headquarters for the bulk of the *Mujaheddin* and camps for Afghan refugees are located. These incursions were occasionally carried out in hot pursuit of *Mujaheddin* guerrillas, but often they were attempts to destabilize the NWFP and exert pressure on Pakistan to cut off its support for the resistance. One of the more serious incidents, in June, resulted in the first aerial dog-fight on the border, between four Soviet/Afghan MiGs and two Pakistani F-16s, in which at least one MiG was destroyed. On 4 January 1987 Afghan/Soviet planes bombed a refugee camp in the Pakistani border town of Chitral, killing and injuring a number of refugees.

Alongside such military activities there were frequent reports of subversive operations – said to be carried out by the KGB-run Afghan secret police, KHAD – which included kidnappings of Pakistani army officers, bombings, assassination attempts on leading *Mujaheddin* and refugee figures, and the promotion of unrest and conflict between the *Mujaheddin* and the local population. Confessions by many KHAD agents captured by the *Mujaheddin* and the Pakistani authorities have shown these activities to be part of an elaborate Soviet/PDPA policy of coercion against Pakistan – a policy which Najibullah alluded to in a speech on 31 August, calling on ethnic Pushtuns in both Afghanistan and Pakistan to unite against the *Mujaheddin* and oppose Pakistan's support for them.

Political Efforts

This intensified effort at military pacification and pressure on Pakistan was combined with an all-out effort to realize Gorbachev's second policy objective: solidifying PDPA rule. There were political initiatives of two kinds, one intended to create unity within the PDPA, and the other to win the minds and hearts of Afghans by rallying certain undecided elements to their leadership.

Disunity between and within each of the PDPA's two intensely hostile factions, *Parcham* and *Khalq*, has from the start hampered Soviet efforts to create a viable PDPA government in Kabul. Shortly after Gorbachev assumed power it became increasingly clear that the

Kremlin wanted to overhaul the PDPA. Since there seemed little real chance of uniting the Party's factions, it embarked on building a two-tier system of Communist rule in Afghanistan. On one level, the PDPA would be maintained as the formal ruling body, manipulated by Moscow as necessary to popularize Communist rule. On the other, the USSR would build up an elaborate secret police apparatus which could serve as the operative mechanism of Soviet control. The KGB made substantial progress in developing the KHAD, which is said to have at present some 20,000 full members and 40,000 informers. Hundreds more Afghans, many of school age, receive training in the USSR as potential future cadres.

The man who assisted the Soviet Union in this task was the chief of KHAD since the Soviet invasion, Dr Najibullah. He proved to be not only totally dedicated to Soviet goals, but also a shrewd, cunning and brutal tactician. Unlike most of the *Parchami* leaders, including Karmal, he is a Pushtun – a member of the largest ethnic group, whose support has always been crucial to any Afghan government's survival. In consequence he was regarded by the Kremlin as someone well equipped to bolster a KHAD-based PDPA rule, free the Soviet Union of the liability of Karmal, and improve the PDPA's acceptability, especially in Pakistan.

Thus, against the backdrop of intensifying pacification operations in late 1985, the USSR set out to replace Karmal with Najibullah. On 21 November 1985 the latter was promoted to the secretariat of the PDPA. In late February 1986 Karmal, Prime Minister Keshtmand (at the time another contender for leadership) and Najibullah attended the 27th Congress of the Soviet Communist Party in Moscow. There they were informed of Gorbachev's new approach to the Afghanistan problem, and Karmal was given a cold reception, in contrast to that accorded to Najibullah. Despite Karmal's best efforts to save his leadership, his days were clearly numbered. On 4 May, during a three-day plenum of the PDPA's Central Committee, Najibullah replaced him as the Party's General Secretary. Karmal remained President of the Revolutionary Council until November, but when Najibullah subsequently consolidated his leadership he was stripped of all Party positions and joined the ranks of such forgotten comrades as his predecessors Hafizullah Amin and Nur Mohammed Taraki.

Pinning its hopes on Najibullah as one of its few indigenous committed Communists, the Kremlin redoubled its efforts to sell him to the Afghan public, and indeed the outside world, and to rejuvenate the PDPA under his leadership. In a series of speeches during May and June Najibullah publicly admitted the PDPA's shortcomings. He criticized its failures to expand its territorial control beyond 'one-third of the country'; to eliminate factionalism and corruption; to bolster the 'depleted' Afghan Army; and to win mass support. This was followed by a vigorous drive to tighten the conscription laws, together with a purge of 'undesirable elements' in the Party (who turned out to be mainly *Khalqis* and the *Parchami* supporters of

Karmal) and their replacement with *Parchamis* faithful to Najibullah.

In mid-July, in keeping with the Soviet political strategy, Najibullah announced the expansion of the PDPA Central Committee from 79 to 147, to include more representatives from the provinces. On 27 September he revealed plans to establish a National Reconciliation Commission to 'normalize' the situation in Afghanistan, and on 4 October it was claimed that the first stage of 'local elections' had been 'concluded successfully in all provinces' (despite the regime's admission that it did not control more than one-third of the country). To give the appearance that Najibullah was winning public support and the situation was stabilizing in Afghanistan, there were more frequent reports of his meetings with ethnic and tribal (particularly Pushtun) leaders. More important, it was announced that six Soviet regiments would be withdrawn in October, and, in a blaze of publicity, these regiments were indeed pulled out.

The Position of the *Mujaheddin*

For most of 1986 these efforts gave Moscow the political initiative, both inside and outside Afghanistan. By the end of the year, however, Gorbachev's revision of the USSR's Afghan strategy appeared to have reached a dead end. Neither intensified military pacification operations nor the attempted revival of the PDPA had paid many dividends.

On the military front, the year proved to be a much better one for the *Mujaheddin* than the year before. They managed not only to stand up to the intensified Soviet/PDPA military operations, but also to launch a number of successful offensives at greater cost to the enemy than before. Several important factors accounted for this.

Although it has been a slow and painful process, the *Mujaheddin* seem at last to have recognized the necessity for greater co-operation. Moreover, they have continued to enjoy effective support from their agents who have infiltrated the PDPA's frail and factionalized administration at all levels (even the KHAD is reportedly riddled with *Mujaheddin* moles). They have also managed to obtain more arms from abroad in the last year, and the quality of those supplies has improved. Their defence against air attacks has been improved by the receipt of man-portable missiles, including a limited number of the effective *Stinger* from the United States. There are no reliable statistics on foreign support to the *Mujaheddin*, but the US arms assistance during 1986 alone was estimated at $250–300 million.

The Soviet/PDPA strategy of forward attack, without an increase in manpower, has meant greater exposure for the troops involved, higher casualties and a failure to occupy and hold the areas taken from the *Mujaheddin*. The latter's infiltration and supply routes have never been cut for more than a short period. The Soviet operation caused severe food shortages for the *Mujaheddin* in many areas, but they proved able to cope, either by rationing or by importing food from other areas. As usual, it was the civilian population that suffered most.

As the year progressed, therefore, the *Mujaheddin* seemed to become increasingly effective, with a string of successful operations to their credit. They not only held their ground in the countryside and many small-to-medium-size towns but also operated in the major cities, most importantly Kabul, which had never come under as much *Mujaheddin* pressure before. The PDPA officially acknowledged the *Mujaheddin*'s widespread territorial control, but it was left for *Pravda*, in a rare report, to admit the frequent disruption of electricity in Kabul.

The War Goes On

Soviet political initiatives neither brought unity to the PDPA nor bought public support for it. Nor did the highly publicized Soviet withdrawal of six regiments impress the opposition or the international community. Involving one tank and three anti-aircraft regiments that had proved superfluous in conditions of guerrilla warfare and the *Mujaheddin*'s total lack of air power, it resulted in no material reduction in overall Soviet military strength. On the contrary, US Secretary of Defense Caspar Weinberger alleged, the USSR had increased its troop numbers beforehand to make up for the withdrawal.

To counter the Soviet PDPA political moves the *Mujaheddin* stepped up their own campaign for better public relations and greater international recognition of their cause. They regard the cease-fire as a consequence of the failure of Soviet policy under Gorbachev, rather than its success, and as a last attempt to drive a wedge between them and their domestic and international supporters.

There is no doubt that the Kremlin faces a very difficult choice over Afghanistan. If it really wants a viable solution that would allow it to pull out its troops, it must address the cause of the problem – the illegitimacy of the PDPA's rule – rather than its obvious symptoms. This would mean entering direct negotiations with the *Mujaheddin* for a lasting settlement, backed by an international guarantee to safeguard Afghanistan's status as an independent, non-aligned state. No other type of agreement is likely to produce effective results – for, if the *Mujaheddin* are not part of an overall settlement, they are quite capable of wrecking it. Moscow must recognize that any possible Islamic government that might emerge would have to operate under at least the same constraints as pre-Communist Afghan regimes: given the country's long border with the Soviet Union, it would be obliged to pursue reasonable neighbourly relations. And, since the USSR has managed to coexist quite well with Khomeini's Islamic regime in Iran, similar coexistence should be possible with an Islamic regime in Afghanistan. As matters stand at the moment, however, the new Soviet approach to the problem of Afghanistan, although superficially attractive, holds out little prospect of a lasting and stable settlement to the conflict that has been the cause of thousands of deaths and ravaged the country over the past seven years.

DEVELOPMENTS ON THE INDIAN SUBCONTINENT

Conflict, or the prospect of conflict, is part of the post-war history of the subcontinent. Events of the past year or two have shown how swiftly its politics can oscillate between co-operation and rivalry, between constructive diplomacy and combativeness. In 1986 there were continuing upheavals in the Punjab, tensions over Afghanistan, insurgency and bloodshed in Sri Lanka, and unsettled borders in dispute between states of the region or between them and third parties. All these testified to the strength of domestic and international pressures and presented threats of one degree or another to the security of the region. It was also a year marked by political stalemate and an increasing loss of confidence in the ability of government to master the complex problems that these societies face in reaching their accommodation with the twentieth century. But at least in one respect there were hopeful, if tentative, signs of progress towards regional co-operation – through the work of the recently launched South Asian Association for Regional Co-operation (SAARC), composed of India, Bangladesh, Pakistan, Sri Lanka, Nepal, Bhutan and the Maldives.

Pakistan: Military Security versus Political Stability

Virtually all Pakistanis who voice an opinion about international affairs believe that their country has been under nuclear threat since India carried out a nuclear explosion in 1974, and that possession of nuclear weapons is the one factor which could compensate for Pakistan's quantitative military inferiority *vis-à-vis* India. Virtually all Indians, on the other hand, hold that Pakistan is actively building nuclear weapons, or may already have done so in secret. They can certainly point to the fact that the Pakistani military explicitly kept this option open after coming to power in 1977.

The 1981 programme of US military aid to Pakistan, amounting to $3.2 billion, ends in 1987, when another programme, of some $4 billion, is to take its place. This would make Pakistan the recipient of the third largest amount of American support after Israel and Egypt. The Symington Amendment, prohibiting aid to any country producing atomic weapons, must be waived to permit approval of the aid package, and the White House will seek to arrange this, as it has in the past. However, claims (later repudiated) by a Pakistani scientist that his country already possessed a nuclear weapon created considerable stir in early 1987, and are bound to weigh heavily with the US Congress. Indian representatives in Washington have taken the unprecedented step of actively lobbying against approval for the proposed aid to Islamabad. Thus two aspects of Pakistan's security policy may be working at cross purposes: the possibility that Pakistan possesses nuclear weapons may be hazarding its acquisition of military technology from the United States at the very moment when the US is showing a new readiness to consider selling advanced military technology to India.

Whatever his success in strengthening military security, President Zia ul-Haq continues to face persistent problems of political stability. Some commentators have observed that Pakistan is becoming harder for any regime, military or civilian, to govern well. The country is economically and demographically dominated by the Punjab, but there is little doubt that the three less populous provinces – Baluchistan, Sind and the North West Frontier Province (NWFP) – are opposed to a Punjabi-dominated system, and their dissatisfaction is becoming more vocal, if not better organized, than before.

Forty years after independence Pakistan is thus still struggling to find its national identity. Most armies in the world are involved with the defence of their country's borders, some with maintaining their own position *vis-à-vis* civilian rivals, and some with the ideological defence of a cause or a system. Under Zia, Pakistan's army is preoccupied with all three and is reluctant to withdraw entirely from power because it fears that no wholly civilian alternative could operate effectively. On the other hand, the idea of permanent military rule is not welcome to most educated Pakistanis, and has in fact been disavowed by President Zia himself. While parliamentary rule has yet to command respect, the old power brokers – the feudal landlords and tribal leaders – are less powerful than before. Maintaining order is becoming increasingly difficult for a country burdened with ethnic, religious, linguistic and other divisions, where there is a large, mobile population, many civilians are almost as heavily armed as the military, and the traffic in both drugs and arms appears to be a growth industry.

Pakistan and the USSR: The Shadow of Afghanistan

Throughout 1986 Pakistan's relations with the Soviet Union were under continuous strain over the question of Afghanistan. Afghan and Soviet military forces frequently violated Pakistan's borders and air space in attacks against *Mujaheddin* guerrillas, and Pakistan had to cope with the major – and worsening – problem of providing asylum to more than 3 million Afghan refugees, currently the largest concentration of refugees from another country anywhere in the world. In some parts of the NWFP, they outnumber the local inhabitants.

Indirect, or proximity, talks between Pakistan and Afghanistan in Geneva, carried on through a UN intermediary, broke down in August 1986, over the question of determining a timetable for the withdrawal of Soviet troops from Afghanistan. Given Gorbachev's professed desire to put the costly legacy of Afghanistan finally behind him, these discussions – which recommenced in February 1987 – must now be assumed to be a focus of Soviet diplomatic attention. The United States, for its part, continued to give Pakistan clear support for its rejection of direct negotiations with Afghanistan and its refusal to recognize the Kabul regime.

Uneasy Neighbours: Unsettled Frontiers

Tense relations with its neighbours preoccupied Pakistan's foreign policy throughout the year. Although the year began with Pakistan and India agreeing to mediate their dispute over the Siachen Glacier, north of Kashmir, and to carry on further high-level contacts, long-standing suspicions of each other's motives and mutual distrust increasingly made themselves felt as the year progressed. Trade talks foundered, as did negotiations on the Glacier, and Prime Minister Gandhi 'postponed indefinitely' what was to have been a symbolic visit to Pakistan in the first half of the year.

More serious was the friction between the two countries which arose from India's continuing problem of Sikh separatism and its charge that Pakistan was actively supporting Sikh extremists. Unusually large military manoeuvres by both sides on their common border, together with inflammatory media speculation, brought this tension uncomfortably close to breaking point. First of all in mid-summer India, attempting to block infiltration by Sikh terrorists from Pakistan, was reported to be massing troops at the border; then Pakistan, in a departure from its usual training pattern, positioned some 14 of its 17 divisions at the frontier with India. The crisis created by this military manoeuvring was eventually defused in February 1987 in a series of high-level talks at which a partial withdrawal of the troops massed on either side of the border was agreed, earlier pledges 'not to attack each other' were repeated, and both countries engaged to 'exercise maximum restraint and to avoid all provocative actions along the border'. Then, in early March, President Zia made a personal visit to India, ostensibly to watch a test match between the two nations' cricket teams. He took the opportunity for an informal dinner and talks with Rajiv Gandhi in Delhi, an encounter which may yet offer grounds for hope that, after a year of strained relations, military posturing and mutual sniping, both countries will try to restart the efforts under way at the end of 1985 to find a new basis of confidence on which to build a more secure relationship for the future.

Twice during the year the focus of tension shifted to the north, as India found itself again in dispute with China over their still unsettled frontier. In July an incursion by Chinese troops and herdsmen into the territory of Arunachal Pradesh was denounced by India. And in December, after the Indian parliament conferred statehood on this territory, Beijing claimed that this was an illegal act, which put further obstacles in the path of an eventual settlement. So far, however, China does not seem to have felt that further action is called for, and it appears that neither side wants see tension escalating over remote border quarrels they have managed to live with for so long. Meanwhile, it is claimed that India's defence capabilities in key, and hitherto vulnerable, areas along its borders with Pakistan and China have been strengthened and are now well balanced to meet any simultaneous Sino-Pakistani attack – the worst-case scenario from India's point of view.

India: A Sense of Drift

For India 1986 was a year of internal turbulence and of declining authority for its Prime Minister, Rajiv Gandhi. Communal and linguistic riots flared up in Gujarat, Tamil Nadu, Karnataka, Maharashtra and even the traditionally peaceful enclave of Goa, and the central government seemed unable – or found it politically inexpedient – to assert its authority visibly and firmly. Gandhi did, however, make some political deals with leaders and erstwhile political opponents in a number of states (notably with Dr Farooq Abdullah in Kashmir, former insurgents in Mizoram, and leaders of the opposition Congress (S) party in Maharashtra), with a view to establishing a more workable equilibrium in India's complex political system.

The Punjab continued to be plagued by intermittent terrorism and violence which worsened as the year went on. On 29 April a five-member 'panthic committee' announced from the Golden Temple in Amritsar the establishment of a separate Sikh state, Khalistan, whose boundaries were to be decided at a later date. On 30 April para-military forces and commandos retook the temple, against little resistance: one person was killed, a number injured, and some 300 detained for possessing weapons. The memorandum of agreement on the Punjab which had been signed on 24 June 1985 by Rajiv Gandhi and the soon-to-be-assassinated Harchand Singh Longowal, president of the moderate Sikh political party Akali Dal, had specified the establishment of a commission to identify Hindi-speaking villages suitable for return to the Haryana state, in exchange for Chandigarh, the capital of the Punjab, ceasing to be the capital of Haryana as well. During 1986 three successive commissions failed to produce recommendations acceptable to both Punjab and Haryana state assemblies in time to meet the dates stipulated for implementation, and as a result the accord remained inoperative throughout the year. Neither the state nor the national government was able to produce a workable political solution in the face of the separatist violence.

Sri Lanka: Another Bleeding Wound

Despite conciliatory efforts by Prime Minister Gandhi and President Jayewardene of Sri Lanka, the bitter conflict between Tamil insurgents and the Colombo government continued to swell a casualty list which already numbers thousands killed and wounded over the past three years. Agreement was reached on broad lines of autonomy for the Tamil-dominated north of the island, but talks foundered on Tamil demands for inclusion of the eastern district where they are outnumbered by Sinhalese and Muslims. To complicate already difficult negotiations, the separatist movement was itself torn apart by the emergence of new and more militant extremists, notably the Liberation Tigers of Tamil Eelam, whose brutality was condemned even among the Tamils in India, who have historically provided the base and support for the insurgent movement in Sri Lanka.

Jayewardene's attempts to pursue a policy of conciliation were not helped by the efforts of the political opposition, the Sri Lanka Freedom Party under former Prime Minister Bandaranaike, to exploit the situation. Moreover, efforts to deal with the insurgency cannot be considered separately from Sri Lanka's relations with India, whose agreement and support is essential for any solution that may be proposed. Prime Minister Gandhi has encouraged negotiation with the insurgents and is prepared to countenance some form of autonomy in principle, but he must take notice of feeling among the 50 million Indian Tamils in the southern state of Tamil Nadu, and he has warned Jayewardene against the temptation to seek a military solution by invading Tamil areas in the Jaffna peninsula. Under such handicaps, even the most persistent and skilful manoeuvring must offer small hope of an early result. As time passes, and political demands harden, the options for compromise narrow correspondingly.

Relations with the Super-powers

In some compensation for its domestic and regional problems, the Gandhi regime enjoyed two successful goodwill visits from representatives of the super-powers. In mid October, US Secretary of Defense Caspar Weinberger came to discuss the supply of defence-related technology, including engines and equipment for Indian light combat aircraft and advanced electronic systems. Despite continued reservations by the Pentagon, which fears the leaking of high-technology secrets, the United States was keen to test the level of Indian interest, remembering the country's traditional wish to avoid becoming dependent on sources of supply which could later be cut off (as had happened in 1965). Although doubts had just been expressed as to whether the US would make an advanced computer system available, it seemed clear at the end of the year that Washington would after all supply a $12-million Cray XMP computer on unusually liberal terms, and on conditions no different from those imposed on its own close allies. Several other deals for US military equipment were under way both before and after Weinberger's visit, but although these arrangements perhaps broke new ground, Indian commentators were at pains to underline that they did not represent a developing 'military relationship' with the US, as Weinberger had suggested.

Mikhail Gorbachev's highly publicized visit in November undoubtedly attracted the most attention from the Indian media during the year. It would be wrong to assume, however, that the visit set a seal on untroubled relations: the fifteenth anniversary of the Indo-Soviet Treaty of Peace, Friendship and Co-operation passed with a pointed absence of comment in India, and there were discreet but unmistakable hints of concern in Moscow (for example in *Pravda* of 14 August) about Rajiv Gandhi's sympathies for western market economics and his preference for Western technology.

All the same, the visit was an undoubted public relations success for the Soviet leader. He repeated an earlier proposal for an Asian-Pacific security conference (this drew no Indian response) but sidestepped efforts to draw him into statements on perennial Indian problems with Pakistan and China. Parallel discussions with the Soviet Chief-of-Staff, Marshal Sergei Akhromeyev, produced agreement for early delivery of MiG-29 aircraft (not then in service outside the USSR) and the possible provision of equipment to neutralize the AWACS aircraft promised by the United States to Pakistan.

A Weakened Premier

For India (as for Pakistan), however, 1986 proved not to be a year that was notable for its progress in domestic security or political stability. At the same time as a large-scale modernization of Indian forces was planned, it was also thought necessary to set up an elite anti-terror force. A widening trade deficit, persistent ethnic unrest and the chronic infighting which characterizes Indian political life all took on even more importance as Rajiv Gandhi seemed increasingly to be losing public confidence, or at least respect. In seven by-elections to the *Lok Sabha* (Lower House) during the year, his Congress (I) party lost one seat and held the others with reduced majorities. In three state assembly elections on 23 March 1987 it suffered more setbacks. Although it gained a power-sharing role in the northern state of Jammu and Kashmir, it lost a similar role in the southern state of Kerala, and it did particularly poorly in West Bengal, where the Communist Party of India – Marxist (CPM) considerably increased its majority. The Congress (I) party is now in power in only twelve of the country's twenty-three states, and shares power in two others. Summing up the year, the editor of the *Times of India*, wrote of a 'mood of despair in India', while an *Indian Express* editorial warned that 'the country appears to be sliding back to the politics of fragmentation' and was increasingly at the mercy of 'the forces of disintegration'.

SAARC: A Qualified Hope

The second full year in the life of the South Asian Association for Regional Co-operation provided guarded hope that this region, so long bedevilled by conflicts of all kinds, may be moving towards a greater sense of shared interests. The organization had come into existence (and, indeed, could continue) only on the understanding that it would not deal with bilateral or contentious issues, and its meetings at Dhaka in December 1985 and Bangalore in November 1986 inevitably offered little in the way of leadership. Nonetheless, the private consultations between India and Pakistan, and India and Sri Lanka, held during the Bangalore Conference constituted the most important diplomacy being conducted in South Asia at that time. In the conference itself agreement was reached to establish a small secretariat (in Kathmandu) and to provide a modest budget, half of which was to be shared between India and Pakistan. After so many years of conflict and rivalry, it would perhaps be shallow optimism to make too much of this initiative. Yet, even this degree of progress is noteworthy, and it would be unduly pessimistic to dismiss prospects for further regional co-operation in the future.

East Asia

CHINA'S REFORMS AGAIN IN DOUBT

Only a decade has passed since Mao Zedong's death in September 1976, yet the social, economic and political changes made in that time have transformed the face of China. Out of the chaos that the Cultural Revolution created in the last years of his life has grown a more stable, more confident country. The highly centralized command economy, based on the Soviet model, has been scrapped in favour of one more market driven and decentralized, in which entrepeneurial Chinese are encouraged to grow rich, and the lazy and inefficient are no longer protected by an 'iron ricebowl'. Investment from abroad and increased trade with capitalist countries (the famous 'open door') have replaced the autarkic, inward-looking tendencies that characterized the Mao years. In place of rigid ideological controls, scientists and intellectuals are encouraged to develop new ideas and to express them freely, even if within limits. China has truly suffered a sea change into something richer and – for orthodox Communists – stranger.

This brief recitation of the changes that have taken place conceals the difficulties that have been overcome along the way and falsely suggests that the path has been a smooth and direct one. In reality, the restructuring and reform of the economy so far achieved has been opposed by a combination of traditionalists in the country's elite who fear for their privileged life style, Maoists concerned at the loss of their vision of Communist economic purity, and conservative military men who worry that their status has been downgraded. It has been a reform driven from the top and has had to rely on timid, poorly trained lower-level managers with a ten-year gap in their education, unaccustomed to demands for innovation, spirit and choice in the running of their enterprises. As a result, it has moved by fits and starts, every two steps forward matched by one step backwards, or sometimes sideways.

The reforms have been heavily dependent on the vision and political adroitness of Deng Xiaoping, whose pragmatism and willingness to countenance change belies his 82 years. He, and other Chinese leaders, however, are products of their revolutionary past. They do not intend to create a capitalist, democratic society in China but are searching instead for a balance between market forces and socialism: 'socialism with Chinese characteristics' as Deng has expressed it. This effort to modernize the economy without reducing the role of the Party has resulted in tension developing between the two forces, tension which at the end of 1986 gave rise to demands from students for a quicker pace of political reform to match the economic reforms which the Party desires. Conservative opponents of the speed of even the economic reforms were able to exploit fears of the new chaos that the student demonstrations illustrated to force through the replacement of the reform-minded Party General Secretary Hu Yaobang, a new campaign against 'bourgeois liberalization', and a pause in the foward thrust of change. Deng Xiaoping's dream of a China very different from Mao's has once again been called in question.

Cooling the Economic Engine

So long as China's economy was piling success upon success, Deng's new approaches could be insulated against the efforts of the traditionalist conservative forces to block them. Effective as they were, however, the reforms were still very much in a transitional phase. Inevitably the costs were mounting alongside the benefits, and in some respects faster, so that by 1985 difficulties had arisen which required immediate attention. There was a fall in agricultural output, while the rest of the economy overheated. Inflation rose significantly (officially to 8.5%, unofficially to about 11%); stresses developed in the reorganizing banking sector, where credit demands outstripped supply; the transport and distribution network had difficulty coping; shortages of electricity and raw materials were intensified by the rapid development; and corruption flourished.

These were all problems brought about by the extraordinary growth that the reforms had created. But they provided ammunition to the conservative forces which allowed them to mount a campaign that weakened Deng and his reformist lieutenants. At the special national Party Congress in September 1985, Chen Yun, the 81-year-old senior Politburo figure around whom the conservative forces appear to have coalesced, made good use of the growing problems in his sharp criticism of the speed of the reform programme. A short-lived campaign developed attacking 'spiritual pollution' (meaning the less salubrious imports from the West, from jazz and jeans to ideas of capitalism and multi-party democracy). The reformers were put on the defensive.

Even without the criticism, those in the top leadership attempting to put into effect Deng's visionary, but imprecise, directions would have found it necessary to make adjustments. Indeed, adjustments had been made many times during the past seven years. They are justified by Deng's pragmatic belief that things should be tried to see if they succeed, and if they do not other approaches should be tried until the one that succeeds is found. In this case the reformist leadership moved on two fronts to bring much needed improvement to the situation.

In January 1986 a high-level task force under the leadership of one of the reformers, Qiao Shi, was formed to tackle the problem of corruption and economic crime. This had two effects: it guaranteed that the emphasis would be on the 'unhealthy tendencies' themselves, rather than on the reform policies, and it kept the campaign that developed out of the hands of the Central Discipline Inspection

Committee, which is headed by Chen Yun. The task force moved swiftly and efficiently in mounting a large-scale propaganda campaign, accused a number of important figures (including the Minister of Astronautics) of illegal activities and, in line with the motto 'kill one to teach 100', summarily executed criminals in a glare of publicity. While the campaign did defuse the political damage that was being done by corruption, it was not wholly effective (the Minister and his deputy got off with only a fine, and those executed were mostly common criminals and rapists). Corruption is an inevitable concomitant of the relaxation of tight economic controls in a country which has always run on *guanxi* (connections), and it will remain a significant problem until the central organization can develop a large enough force of skilled auditors to ensure control.

The purely economic problems were successfully managed by putting the brakes on the runaway economy. The major action was to restrain excessive lending by state-run banks, mainly through raising the interest rate for loans. To relieve the inflationary pressures prices were frozen. The restriction on the availability of investment funds was buttressed by the natural effect of a shortage of raw materials and energy supplies and led to a dramatic slow-down in the growth of the economy. In 1985 the industrial sector had grown by 23.1%; in 1986 this was cut back to an overall rate of just under 8%. Heavy imports of consumer goods had caused a trade deficit in 1985, notably with Japan, of $14.9 billion and created a foreign currency crisis. In July 1986 the regime devalued the *renminbi* (the Chinese currency unit) by 13.6%; coupled with tighter import controls, foreign-exchange shortages and the slow-down in economic growth, this has begun to bring the trade picture back into balance, with imports peaking in the middle of 1986.

One Step Forwards, Two Steps Backwards

With the economic problems under better control, the reforming leadership turned once again to political activity. Deng Xiaoping seems to recognize that in order to create the proper atmosphere for decentralization of the economy, with its emphasis on local managers making decisions, to work, it would be necessary to reduce somewhat the stultifying centralized controls of the Party. Yet efforts to relax controls and democratize the Party machine run directly against the beliefs, privileges and power of the men who have been running China since the revolution. And while Deng intends to loosen the Party's monopoly of information and allow skilled members of the society to come up with new ideas, even he does not intend the phrase 'socialist democracy' to mean allowing the people to have the final say in how the country should be run.

Earlier efforts to loosen controls somewhat – in the 'Hundred Flowers' campaign of 1956 and the Democracy Wall experiment in 1979 – had both failed. In both cases, the criticisms the Party had called for went beyond providing new ideas on how to invigorate the economy and attacked the Party's centralized control itself; the campaigns were therefore abruptly ended, and the intellectuals who had bravely put their heads above the parapet suffered severely. Deng and the leadership recognized that the current drive for economic liberalization required the controls to be loosened, and they were confident that enough changes had now been made that a new liberalization effort could be carefully controlled; this led them, from the late spring of 1986 onwards, to try once again to allow intellectuals to speak their minds.

On the thirtieth anniversary of the Hundred Flowers campaign, a new call went out for 'letting flowers bloom', particularly in the academies and scientific institutions. The campaign was given added weight by the appointment in June of a widely admired novelist as Minister of Culture and the replacement of Deng Liqun, a leader of the conservative wing of the Party, as Director of its Propaganda Department. Films and novels appeared which were critical of the way Party members used their positions to improve their private lives and to sabotage economic reforms. An extensive discussion of the need for political reform appeared in the Chinese press throughout the summer with the clear backing of the General Secretary, Hu Yaobang.

The discussion was intended to pave the way for a new statement on how economic change and Communist ideology could be correctly balanced, which was to be presented to the sixth plenary session of the Party's Central Committee. When the statement was presented and approved on 28 September it was apparent that the leadership had not yet been able to square this particular circle. The original version had reportedly been changed three times under pressure from conservatives, and they had apparently been able to water down the more liberal aspects of the document. Yet the thrust was still very pro-reform. The drive for academic, literary and artistic freedom was made official Party policy. 'Democratic centralism' was upheld as the abiding rule of Party life, but Marxism was not to be treated as rigid dogma. Pragmatism, ensuring that practice stayed in step with changing realities, was to be the motivating philosophy in place of unchanging *a priori* principles; 'freedom of discussion and freedom of criticism and counter-criticism' was affirmed. And the policy of an opening to the world was reaffirmed.

Encouraged by this stand, increasing criticism of Party control was voiced, particularly by a number of university professors, and the call for greater democracy was taken up by university students in December, the traditional period for student demonstrations. In 1985 such demonstrations had been handled with care by the authorities, and they had not got out of hand. With the new climate, however, the demonstrations – which began in provincial universities in Hefei and Wuhan – soon turned from bread-and-butter demands for better living and working conditions to requests for elections and greater democracy. The authorities again treated the demands carefully; in

Hefei they even postponed the election of representatives to the provincial Party Congress from 8 December until 29 December to allow the students to nominate candidates.

However, when the student protests moved to Shanghai and then to Beijing, were being mounted every day, and involved thousands whose demands for democracy became more strident, then even the liberal leaders in the government became concerned. Their view of the role of 'democracy' was that it should make the ruling Communist Party more responsive to needed change, not lead to outright criticism of the Party itself. While it is possible that the initial student protests had had official encouragement, in the hope of prodding the conservatives along, they had now gone too far, and the authorities tried to restrain them: first by condemning them in the press and then by bans on further demonstrations coupled with gentle police action. Neither worked, however. Not only did the students defy the bans on demonstrations, but on 5 January 1987 in Beijing they burned copies of the *Beijing Daily*, which they claimed had distorted their protests. These acts, reminiscent of the actions of the Red Guards during the Cultural Revolution, alarmed the authorities and embarrassed the liberal leaders. The month-long student campaign, growing stronger and more radical as it went on, had by now brought conservative voices out in force.

Thrown on the defensive, Deng Xiaoping moved swiftly to try to repair the damage. On 8 January he called for the removal of 'big shots' in the Party who were arguing for Western-style democracy and condemned in particular Professor Fang Lizhi, who had encouraged the first student demonstrations in Hefei. Fang was dismissed from his post as Vice-President of the University of Science and Technology of China and disappeared from sight. On 13 January the Director of the Information Bureau of the Party's Propaganda Department was sacked. And on 17 January General Secretary Hu Yaobang, the most outspoken of the liberal leaders, resigned. His position was taken by Premier Zhao Ziyang on a 'temporary' basis; Zhao also kept his post as Premier, suggesting that he will fill both positions until a new Premier is named.

Since that dramatic change, events in China have indicated that the reforming drive has been curtailed. The head of the Propaganda Department of the Party was replaced on 4 February by a hard-liner. Although Zhao had insisted that whatever clean-up was required would be restricted to the Party ranks, the Army joined in the anti-liberalization drive in the middle of February, calling for a campaign of ideological indoctrination for its own ranks. At the end of February, when students who had been on holiday for the Chinese New Year returned to school, they were greeted by the republication of a speech that Deng Xiaoping had made in 1962 admitting Party mistakes, praising democratic centralism and attacking 'bourgeois liberalism' – the present code name for capitalism and Western-style democracy.

Both Zhao and Deng have been insisting at every opportunity that none of the recent developments means that the economic reforms are in jeopardy. It is difficult to believe, however, that there will not be residual effects; at the least the reforming climate which underpinned Deng's efforts to invigorate the economy will be chilled. Those elements in the Party (and there are many) who have felt that factory managers, writers and intellectuals were being given too much autonomy will now find it easy to condemn actions they do not like as representing 'bourgeois liberalism'. The position of the liberals must have been weakened, and many throughout the country will be unwilling once again to thrust themselves forward with new ideas and activist policies until the situation is clarified. Deng's task, now that he has been forced to act to ensure stability and order, is to see that it does not once again regress into rigidity.

A Steady Foreign Policy

There has been no indication that foreign-policy issues have played any role in the differences between the two factions in China. For many years there has apparently been agreement on a policy of balancing carefully between the two super-powers, while abjuring any activist policies in the rest of the world. The leadership has been concentrating on domestic developments and appears to believe that they can best be pursued in a peaceful environment; most of its moves in the foreign arena have been dedicated to this end. During 1986 this policy was pursued, and there is no sign that it will not be continued even while the internal political struggle goes on.

China's resolve to maintain a balance between the two superpowers was tested in 1986 as Gorbachev pressed a campaign for a quickening and deepening of relations between the Soviet Union and China. In his seminal speech at Vladivostok in July Gorbachev made gestures intended to smooth the three 'obstacles' which China insists must be removed before relations can become closer. The proposed withdrawal of 6,000 troops from Afghanistan, a promise to draw down Soviet troops in Mongolia, and a suggestion that the outstanding problem of the Amur river border could be settled by drawing the border through 'the main shipping channel' (which would help the Chinese claims to some of the river islands) were all intended to entice China to be more responsive. Appointing Oleg Troyanovski as Soviet Ambassador and replacing Leonid Ilyichev with Vice-Foreign Minister Rogachev for the ninth round of bilateral consultations between the two countries in October were other signs of Soviet seriousness. Finally, a suggestion by the USSR that it would welcome a meeting between Gorbachev and Chinese leaders (presumably Deng Xiaoping) capped the Soviet campaign.

China remained wary of Soviet enticements. Deng Xiaoping agreed that there was some apparent change in Soviet attitudes, but continued to insist on the central importance of the third 'obstacle', Soviet aid to Vietnam in Kampuchea. Since Gorbachev had not dealt with this in his Vladivostok speech it effectively put brakes on the Soviet drive. Deng Xiaoping also turned down the idea of a summit meeting, saying that he would be prepared to meet Gorbachev only if some real advance on the Kampuchean question were made. Yet Beijing welcomed the move on the border question, and in February 1987 the two sides met in Moscow to re-open talks on this problem that had last been held in 1978. There were other signs of slow improvement in Sino-Soviet relations: among them exchanges of visits at ministerial level, an increase in trade, and an agreement, signed at the end of a visit to Beijing by Soviet Deputy Prime Minister Ivan Arkhipov, for Soviet engineers and technicians to help in modernizing 17 of the industrial installations that the USSR had built in China in the 1950s.

China took the same careful line with the United States. A number of high-level US officials visited Beijing, including Treasury Secretary James Baker in May, Secretary of Defense Caspar Weinberger in October, and Secretary of State George Shultz in March 1987. Each of these meetings advanced one aspect or another of Sino-US relations: Baker's visit helped US firms doing business with China (revised regulations for foreign investments were promulgated in October), and at the end of Weinberger's trip the first visit to China by US warships was announced (one in 1985 had been stymied by Hu Yaobang's offhand remark that they would have to declare that they were not carrying nuclear weapons). During the year the US decided to proceed with a long-outstanding sale of military technology, including avionics for Chinese fighters and submarine torpedoes. Yet China made clear that the US relationship with, and sale of military equipment to, Taiwan was still seen as an insurmountable obstacle to closer relations. The careful balancing act between the super-powers appears likely to continue for some time.

Slow Ahead

Whether Deng and his reforming lieutenants can recover from the blows they suffered at the end of 1986 is now an open question. Although Hu Yaobang was a protégé of Deng's and had been put forward for the General Secretary's position by Deng, there were indications before his dismissal that the two did not wholly agree on the pace at which political liberalization was being pushed in China. If this is true, his removal as General Secretary may not be quite the setback to Deng's influence and plans that it now appears. Yet the conservatives have clearly been strengthened in their desire to slow the changes and to ensure that centralized Party control remains the guiding rule in China.

In the past two years Deng Xiaoping has made many references to his desire to retire, leaving the running of the country to younger men – men, of course, who would share his views. At each turn, however, it has been evident that the security of his policies was still in doubt, and he has been 'convinced' by others of the need to stay at the top. But he cannot last forever, and it has been clear that the Party Congress due to be held in October 1987 is the point at which he wishes to turn the reins of power over to others whom he can trust to see his reforms through.

The time left before the Party Congress has clearly become a key period that will help decide whether the extraordinary effort to transform China's economy and society will continue to be pursued, or whether the leadership will return to a more orthodox Communist model. What looks like a classic succession struggle is now being waged behind the scenes in China. If the upward turn of the economy falters, or if Deng is forced to relinquish his control, either through illness or death, it seems doubtful that those who would continue the reforms will succeed. But both Deng and the economy have shown remarkable resilience. The odds must still favour a steady, but slower, march along the path on which he set China after Mao's death.

JAPAN: AN UNEASY YEAR

In July 1986 the Japanese electorate gave Prime Minister Nakasone and his Liberal Democratic Party (LDP) an overwhelming electoral victory. The LDP won 300 of the 512 seats in the Lower House of the Diet and 142 of the 251 seats that were up for election in the Upper House. This outcome was widely recognized as a personal triumph for the prime minister, who had overcome resistance by leading members of the party to his plan to holding simultaneous elections to both houses of the Diet. It was largely due to this double poll, which brought the voters out in unusually high numbers, that the LDP was able to reverse its abysmal showing in the December 1983 election, in which it failed to achieve a majority and was only able to govern by attracting votes from smaller parties. Nakasone's daring was rewarded when, in the face of this success, the other LDP leaders felt themselves constrained (albeit reluctantly) to agree that he should stay in office for one more year – despite the party's regulation which limits its president (*ergo* the prime minister, when the party is in power) to four years in office.

Nakasone may have been hoping for his term to be extended for two years, but the other factions within the LDP, who wish to see their own men in power, were certainly not willing to go that far. Indeed, the results of the election probably took them by surprise; the so-called 'New Leaders' – Noboru Takeshita, Shintaro Abe and Kiichi Miyazawa – had spent much of the first part of 1986 manoeuvring to secure a good position for the leadership election that had been due in October 1986, at the end of Nakasone's second two-year term as party leader. They will now have their sights fixed on October 1987, when his reprieve runs out.

But since the apogee of his popularity in July, Nakasone and his government have dropped precipitously in the people's favour, and the 'new leaders' may be pleased that he has been around to take the blame. Part of the problem has stemmed from ill-considered statements by Nakasone himself and one of his ministers, but a more significant part derives from the prime minister's attempt to tackle such unpopular questions as tax reform and the privatization of the Japanese National Railways (JNR).

Soon after the LDP victory the newly appointed education minister, Masayuki Fujio, publicized his support of the rationalization of Japan's role in World War II in revisionist Japanese history text books, which had led to strong protests from China and Korea in 1985. In an interview he went much further, stating baldly, among other things, that Korea had been 'occupied' by Japan in 1910 by agreement between the two countries' leaders. The outcry from Korea stung the prime minister into action; when Fujio refused to resign Nakasone sacked him and ten days later apologized. Although this quieted foreign dismay, the incident disclosed how much support Fujio's beliefs had in some sections of the LDP, and it could only reinforce the views of those elsewhere in the Far East who remain suspicious of Japan.

Nakasone's action in this affair, although perhaps somewhat tardy, did not rebound directly against him. But his own remarks to an LDP meeting in September, seeming to disparage the low mentality of minority groups in the United States and compare this with the higher intelligence of 'pure' Japanese, were another matter. In the end Nakasone was forced to apologize to the American people for his remarks (which he claimed were misjudged and taken out of context), and then to the Japanese public for the embarrassment he had caused them.

In November, despite the strong opposition of the other parties, the government pushed through the Lower House a bill to privatize the JNR. This will split the railway into six regional entities, cause up to 60,000 workers to be laid off, and require the sale of railway real estate to pay the old debts. None of this improved Nakasone's standing with the people.

But more serious for the LDP and Nakasone were the moves made at the year-end to determine the party's final position on implementing a significant tax reform. In October the prime minister's advisory board recommended cuts in personal and corporate taxes, suggesting that the resultant revenue reductions should be met by new indirect taxes and a tax on the hitherto tax-exempt small bank deposits, called *maruyu* in Japanese. These recommendations were then taken into the deliberations of the LDP Tax Commission, which was working to a deadline of 1 April 1987. This is a very powerful body, being made up of serving Diet members from the ruling party itself, which largely drafts new tax legislation for consideration in the Diet. It has already embraced the idea of taxing *maruyu* at a rate of approximately 20% on the interest they earn and has also proposed a modified value-added tax as a way of raising some $18 billion.

The final shape of a tax reform bill will not be known until later in 1987, and may not be decided even then. However, the opposition parties have made their unhappiness clear, and even many members of the LDP are having second thoughts about the bill. Demonstrations have been held against the reforms (at one of them an effigy of Nakasone was battered by passers-by), particularly the notion of introducing value-added tax, and it seems certain that the tax reform problem will affect the LDP's chances in the forthcoming local elections.

The International Sphere

Soviet–Japanese relations, which at the outset of 1986 had looked as if they were about to be considerably improved, ended the year in much the same state as they have been for some time. In January Eduard Shevardnadze made the first visit to Japan by a Soviet Foreign Minister in ten years, and it seemed that some agreement on the resumption of negotiations on a peace treaty was reached. However, during foreign minister Abe's visit to Moscow in May it became clear that for the Soviet Union such negotiations would not include discussion of the issue of the disputed Northern Territories which has been the stumbling block to any improvement in Soviet–Japanese relations. While Abe did sign a cultural exchange agreement that had been under negotiation for fifteen years, the long-term economic package that the Soviet Union had wanted was not agreed.

Japan also reacted coolly to Gorbachev's Vladivostok speech setting forth Moscow's ideas for rapprochement with all its Asian neighbours. Nor was it willing to accept a proposal that Gorbachev should visit Japan for talks; much as Nakasone would like to see him in Tokyo (for the increase in prestige it would bring him), unless the Soviet leader were prepared for serious discussion of the Northern Territories question, he would not be welcome. Failing some give in the Soviet position on the Northern Territories problem, relations between the two countries will continue to be restricted.

Education Minister Fujio's remarks had thrown doubt on the prime minister's planned trip to Korea, and a visit by Crown Prince Akihito and his wife was cancelled. But after Nakasone had apologized he visited Seoul in September for the opening of the Asian Games. Not many concrete advances were made during this trip, but Nakasone did promise to make changes in the Japanese Alien Registration Law, so that Koreans living in Japan would only have to be fingerprinted once. He also promised to send a trade delegation to look into the possibility of increasing Japanese imports from Korea so as to reduce the heavy imbalance in trade between the two countries.

Nakasone also did some fence mending during an October trip to China, which had also been incensed by Japanese insensitivity on the text book revision issue and renewed visits by Japanese politicians to the Yasukuni shrine (a Shinto war memorial which Nakasone had officially visited in August 1985). Japan would like closer security relations with China, and during a visit by the Chief-of-Staff of the Chinese armed forces in May some closer co-operation between the two defence establishments had been agreed. Nakasone's visit does not seem to have advanced this much, but a visit to Beijing by the Japanese Defence Agency head, planned for April 1987, may see some changes.

Economics and Trade

The dynamic leadership that Prime Minister Nakasone had shown in the political field had extended also into the economic and trade spheres, with results that were impressive by any standard. But difficulties caused by a sharp rise in the yen against the dollar now darkened the picture. On 24 January 1986 the yen closed at 195 to the dollar; almost exactly twelve months later, on 19 January 1987, it closed at 149. In September 1985 the rate had been 240.

January 1986 yen rates in the 190s had set off serious 'rationalization' pains among many small to medium exporters in Japan, and the government responded with special relief measures that caused an almost immediate outcry from US and EEC critics. The '*endaka*' crisis (as yen appreciation is called in Japanese) and concern over the growth of protectionist sentiment in export markets continued throughout the year, causing Japanese firms to consider exporting jobs to Korea, Taiwan, the ASEAN countries, the EEC or the United States. By the end of the year, however, the Tokyo government was trying to maintain a stable exchange rate at around 160 yen to the dollar. Failure to control the yen's surge and the dollar's decline in the wake of an October bilateral exchange rate agreement with the US was reportedly causing some resentment among Japanese business and financial circles. This was reciprocated by some US dissatisfaction that Japan's FY 1987 budget fell far short of the expansionary level needed to re-ignite the country's economy – which only realized 3.5% real growth during 1986.

In April one of Prime Minister Nakasone's *ad hoc* advisory groups produced a significant report. The recommendations of the Advisory Group on Economic Structural Adjustment for International Harmony – the *Maekawa* Report – called for Japan to energize future economic prosperity by focusing inwards upon domestic markets. Recognizing that Japan could not forever continue to depend on an export-oriented economy and expansion of its market share overseas, the report called for a fundamental restructuring of the economy. The obsolescent tax system and the obsolete agricultural system (which leads to Japanese rice costing five times more than the world price) came in for some close attention, and tax shelters for small savings and over-protected farmers were put under notice. The report, which strongly recommended measures to realize an international division of labour throughout the Western world, was endorsed by the Prime Minister's Economic Council shortly after the May economic summit. Nonetheless, Japan's overall trade surplus for 1986 reached a level approaching $80 billion – out of which the surplus with the US was over $50 billion. Moreover, the appreciation of the yen plus the slow-down in the economy cut imports of raw materials in the early months of 1987, which drove the surplus into greater imbalance.

In May Japan hosted the seven-nation Economic Summit and drew praise from President Reagan for the most successful such summit to date. But Nakasone came in for domestic criticism for endorsing the summit's anti-Libyan declaration and for not achieving a general agreement to halt the continuing depreciation of the dollar.

Several issues continued to gnaw at economic and trade relationships with Japan's trading partners. One seemed to have been solved, but others continued to evade rapid solution. A dispute with the US over production and sales of semiconductors was successfully resolved as July came to an end. By October, however, although Japan was no longer selling 256-kilobyte and DRAM (dynamic random-access memory) chips at cut rates, the prices had gone so high and sales had dropped so much (because consumers had accelerated their orders before the price increases) that both US and Japanese producers were having problems trying to even out supply and demand.

One of the current disputes affecting both EEC and US interests is the $8-billion Kansai Airport Project being built on reclaimed land in Osaka Harbour. In order to compete with Japanese firms, foreign companies must have a 'proven track record in Japan', which means that foreign construction firms without a long history in Japan are denied access to this attractive and possibly overpriced endeavour. With Japanese firms having done some $2 billion of business in the United States during 1985, US representatives are getting less and less patient.

Efforts to respond to American desires for the stimulation of Japanese domestic demand were announced in mid-September. The Nakasone government allocate some $23 billion under a stimulus package: $14.8 billion for public works projects, $4.5 billion for public housing, $2.3 billion for utilities, and significant amounts for communications ventures and loans for small and medium-sized firms in difficulties.

Defence and Security Affairs

1986 will be remembered as the year in which the ten-year-old defence expenditure ceiling of 1% of GNP was finally abandoned. At the end of 1985 Japan's defence spending had reached the point where Nakasone's firm commitment to fulfil the goals of the 1986–1990 mid-term defence programme had become impossible unless the 1% ceiling were to be abandoned. In past years the limit had been adhered to despite rises in annual defence expenditure because the country's GNP had risen as well. But, with the advent of slower growth, an increase in spending was called for if the new defence plan was to be fulfilled.

The abandonment of the 1976 policy does not mean that the floodgates have been opened to limitless increases in Japanese defence expenditure, although it is clear that future defence budgets will go beyond the 0.004% advance on the 1% ceiling approved on 24 January. The LDP did not establish any new quantitative limit on defence spending but said that, although moving beyond 1%, it would continue to respect the 'spirit of the ceiling of about 1%'. But as the years of the new defence plan go on, it will become more and more difficult to stay close to 1%, as was achieved in 1986, because growth in the GNP cannot be expected to create the extra funds needed to keep the plan on schedule. Nonetheless, the rise will not be precipitate; the pacific mood of the Japanese, the tight situation of the national budget and the negative reaction of regional neighbours will all work against this.

Another significant step forward in the defence and security area was the creation of an intra-governmental co-ordination system in the Cabinet Secretariat. The creation of the External Affairs Office and the Internal Affairs Office within the secretariat will enhance policy co-ordination amongst ministries and agencies, and greater top-down guidance will be possible. Such a step was becoming necessary because the traditional bottom-up decision-making was increasingly ineffective in a trade and defence environment that requires responses worthy of the computer age.

The crisis management system was also overhauled, with the National Defense Council being scrapped and replaced with a National Security Council. The new body will have the old defence policy approval responsibilities of the NDC, but it will also manage government responses to 'political, economic, social and military contingencies affecting national security'. Many saw in this reorganization an attempt by Nakasone to create a modified prime ministerial system of government, giving him greater power to affect policy formulation than the traditional system affords. However, such modifications to the foreign-policy and trade apparatus had been recommended by the Commission on Administrative Reform.

Besides these decisions affecting the structure and substance of Japanese defence and security policy, another very important step was taken in September, when Japan announced its willingness to conclude an agreement with the US on participation in SDI research. After a series of meetings held over several years, Nakasone's 'understanding' of President Reagan's SDI concept was accepted by both sides. However, it can be inferred from the fact that meetings have been held since September without any agreement being announced that difficulties have arisen in implementing this decision. The exact nature of participation will be determined during a further series of meetings continuing into 1987.

The FY 1987 defence budget announced in the closing days of 1986, besides breaching the 1% ceiling, included a 5.2% increase in defence spending, rather than the 4.1% proposed by the Finance Ministry. While not the 6.7% increase requested by the Japanese Defense Agency, this would permit the JDA to procure a considerable amount of additional equipment, so that the second year of the 1986–90 five-year plan could stay on schedule. With this rate of achievement of the plan (about 20% per year), Japan will be able to fulfil the 1976 National Defense Program Outline (*Taiko*) by 1990 or so.

Other defence- or security-related developments included: the start of an in-depth self-examination by the JDA that identified 32 items to be studied (at the end of 1986 six major internal reforms were being staffed); a continuation of 'look north' defence planning to restructure the Ground Self-Defense Force so that in a crisis a 100,000-man force, highly mobile and with significant firepower, would meet any Soviet aggression in Hokkaido; an increased Japanese government commitment to support US forces in Japan caught by the sharply increased value of the yen; completion of the joint US–Japan sea lane defence study; the holding of the first US–Japan joint military exercise off Hokkaido to involve the combined services of both countries; and the election of Japan as a non-permanent member of the United Nations Security Council.

Prospects

Under Prime Minister Nakasone Japan has begun to play a more energetic role in respect of security, foreign policy and foreign aid. It has become somewhat more difficult for critics, particularly in the US, to fault the country for unwillingness to make proper arrangements for effective security against a possible enemy. However, the domestic difficulties that now confront the prime minister, particularly the tax increase, make it questionable whether he will be allowed yet another year of rule. And it looks likely that the strong position that the LDP achieved in the summer of 1986, partly as a result of his personal popularity, will be eroded during 1987. Nakasone's ideas of how to balance Japan's unquestioned economic weight with a more positive role in world affairs will probably not be pursued by him or whoever might succeed him with the same degree of energy that he has hitherto invested in them.

THE KOREAN PENINSULA

Politics in South Korea have always been volatile, and 1986 was no exception. Attempts to chart an acceptable route towards a more democratic government have been hampered by the lack of trust between the present government and the opposition forces and by the development of more extreme views on both sides. The stakes in the struggle are rising and time for solving the problem is running out.

President Chun Doo Hwan's term of office will end in Spring 1988, shortly before the summer Olympics are due to be held in Seoul. All South Koreans would like an agreed solution to the transition of leadership to be decided before the Games, for it is hoped that they will bring international recognition that South Korean society and government have become mature and stable, with the people supporting a popular government that is effectively advancing their economic and political interests. The major stumbling block remains the definition of what is an effective and stable government.

In North Korea, where government actions are shrouded by a complete lack of information, politics have appeared rigidly settled. But there were indications in 1986 that, even there, the problem of a transition of power may have begun to disturb the calm. Kim Il Sung has been the supreme leader of the country since the end of World War II, and his rule is not in question. But his desire to see the baton of power handed on to his son may have caused an unexplained flurry of political activity and rumours of possible efforts to thwart his wishes. If so, they clearly did not work; Kim is still very much in control, and his dynastic plans appear to be moving forward.

The Violence of South Korean Politics

If President Chun Doo Hwan should step down from his position at the end of his constitutional term, and power were to pass peacefully to his successor, it would be the first time since the republic was established in 1948 that a leader had not been replaced by coup or assassination. President Chun has promised to do just that, and it is probable that he plans to do so. But his promises did not satisfy the opposition, which claimed that power would simply be transferred to another general, and the army would still be the ruling class. It insisted that the only way to prevent this would be to amend the constitution, instituting direct democratic election of the President in place of the present indirect election by an appointed electoral college.

At the beginning of 1986 the opposition New Korea Democratic Party (NKDP) announced that it would mount a campaign to gain five million signatures backing a constitutional change, and it initiated a series of mass rallies. President Chun countered by threatening to arrest the participants, and a sense of crisis gripped the country. In February, however, came news of the peaceful 'people's revolution' in the Philippines. The situations in the two countries were quite different – one an economically failing country, ruled by a corrupt autocrat with minimal support; the other economically growing, with a ruler who, if autocratic, was not corrupt and was backed by a strong and cohesive military machine – but there were enough resonances to make Chun drop his threat and reconsider his position. Whether or not he was shaken by the fall of Marcos and by delicate US pressure to make some adjustments to ensure that the same did not happen to him, the opposition was vastly encouraged.

But the mass demonstrations that sprang up throughout the country were taken over by more radical elements among students and workers and acquired an anti-American and anti-imperialist cast. In May a particularly violent street demonstration in Inchon was met by firm police action, and the initiative shifted in favour of the government. There is indeed a considerable desire amongst South Korea's newly-developed middle classes for more democracy and an opportunity to participate more fully in the government process, but there is also a fear of radical solutions. And, since radical extremism tends to be attributed to Communists and agents of the North, the government's repression of violent demonstrations could now be viewed as protection against that ever-present threat.

The open appearance of radicals also upset the uneasy coalition that makes up the opposition. Unlike the ruling Democratic Justice Party (DJP), which is firmly united on its goals, the opposition is united only in wanting to remove the current military-based government, and each of its elements has different ambitions for the future. In March five major organizations had banded together in the National Alliance for Constitutional Reform to press the signature campaign. Six leaders were named to its council: Kim Dae Jung and Kim Young Sam of the Council for the Promotion of Democracy (currently the strongest leaders in the NKDP); Lee Min Woo, representing the NKDP, of which he is the nominal leader; the Rev. Moon Ik Hwan, president of the United *Minyung* (People's) Movement for Democracy and Unification (UMDA); and two churchmen representing the Korea National Council of Churches (KNCC) and the Korean Catholic Church. Under the new pressures, this unwieldy and hastily constructed coalition began to crack.

Kim Dae Jung is an ambitious and popular politician who, even by the official count, had come close to winning the last open election in South Korea in 1971; many believe that he had in fact won, but for subsequent vote-rigging and false counting. In 1973 he was seized by Korean agents while on a speaking tour in Tokyo, returned to Korea by force and kept under arrest until 1980, when he was accused of pro-Communist beliefs and condemned to death. He was only saved by US intervention. After a period of asylum in the United States he returned to Korea, and his sentence was suspended; even so, he is prevented from taking part in any election that might be held, because his sentence was not abrogated. Kim Dae Jung is pro-American and anti-Communist, but somewhat to the left of his colleague and rival Kim Young Sam, a more centrist politician; many believe that, if standing alone against a DJP candidate in a free direct election, either could easily win. Lee Min Woo, more rightist than either, is a rather colourless figure with none of the personal charisma of the two Kims; he would be more willing than either of them to reach some kind of compromise with the governing party.

The Rev. Moon Ik Hwan and the UMDA which he leads are far to the left, anti-American and bitterly opposed to the military government. In March the party issued a statement calling the government 'a military-dictatorial regime', and in May, as a result of anti-government statements, Moon was arrested and the rest of the leadership of the party went underground. In October Moon, who has spent much of his life in jail, was sentenced to three years imprisonment.

The government was able to split the opposition further when an agreement was reached between Lee Min Woo and Roh Tae Woo, DJP chairman and a former general, who is widely thought to be the DJP's choice for the country's leader after Chun steps down. Lee and Roh had both been in Washington in the middle of May and had had talks with Administration officials, including Secretary of State Shultz. The accord reached by the two men called for a special session of the National Assembly to examine the possibilities for a constitutional change that would ensure the people's choice in the government. The NKDP had hitherto insisted that no agreements with the government could be reached until political prisoners were released from jail and the rights of those, including Kim Dae Jung, who were barred from political activity were restored. Lee's deal with Roh thus made plain the splits that had developed in the opposition; and, since Lee is a member of the faction led by Kim Young Sam, the differences seem to extend to the NKDP's two most important factions and their leaders.

With the formation in June of a special committee of the National Assembly to prepare the way for the special session, the government unveiled a new plan for governance. Hitherto the DJP had insisted that no change in the constitution would be possible until after the elections in 1988 – elections which, if held under the old rules, would have ensured a President from the DJP. The new plan, announced in mid-August, called for a complete change in the governing structure. It would reduce the President to a largely ceremonial role as head of state, transfer most of his present powers to the Prime Minister (who at present has little power), and both would be elected by the National Assembly. Although there were a large number of important details left to work out, the ruling party hoped that the Assembly could pass a law along these lines by the end of 1986. The constitution requires that an amendment of this kind must then go to the people in a national referendum; if this could take place in 1987, the way would be cleared for elections to the National Assembly in 1988 when Chun's term ends.

The NKDP, however, refused to make any changes in its demands for direct election of the President and threatened to walk out of any special session which passed any other amendment. During the remainder of the year, however, the strains in the party became more apparent. Kim Dae Jung declared that he would not run for President if there were direct elections, but this proclamation was greeted with some scepticism. Near the end of the year Lee Min Woo issued a seven-point plan to secure a number of civil rights and said that, if the government would accept them, he was prepared to discuss a cabinet, as opposed to a presidential, form of government as the basis for a constitutional amendment.

Bolstered by the seeming disarray in the opposition, the government moved to prevent street and student demonstrations from getting out of hand. In October large numbers of police entered Seoul National University, where students had been demonstrating for days, and arrested over 1,200. At the end of November over 70,000 riot police sealed off central Seoul to prevent a rally, called by the NKDP to support its proposals for direct elections, which it had expected up to a million to attend. In the event there were more armed and shielded police than demonstrators; because the student demonstrations seemed to have taken on a more radical hue, the government's repression of dissent did not create the kind of support the opposition had hoped for.

Just as it appeared that the government was on top of the situation, though, the initiative slipped away from it again. In early January 1987 it was revealed that an arrested student had died under torture in police custody. The government moved with praiseworthy speed and candour, mounting an immediate investigation, exposing all the embarrassing details and then arresting two officers and charging them in effect with murder; the next day Chun sacked his Minister of Interior and chief of police, expressed his own regret at the incident and established a special agency to protect human rights. Despite these actions, however, the damage has been considerable. The opposition has now reknit its coalition, with Lee Min Woo withdrawing his seven-point proposal and the NKDP as a whole reaffirming an insistence on direct presidential elections. Its firm stand on the torture question has gained it new support from religious groups and human-rights activists, and many believe that it will now be able to reconsolidate its relationship with all but the most radical students. The government is once again on the defensive, with time running out for a solution before the preparations for the 1988 summer Olympics must be made. Unless agreement can be reached on the constitutional question, more violent demonstrations and firm police reaction can be expected throughout 1987. In that event, the image of South Korea that will strike foreigners when they arrive for the Games will not be the one of a mature and stable society that Chun had hoped to present, but that of a fortress under siege.

The North Korean Succession

If South Korean politics are volatile and largely unpredictable, the same has not been true of North Korean politics in the past. Kim Il Sung, leader of the country ever since its establishment in 1948, has seemed to be unchallenged for almost as many years. The cult of personality which surrounds him is extraordinary, even for Communist countries, and he has also appeared to have the support of the military and key Party members. Since at least 1980, when the sixth Korean Workers' Party Congress legitimized his plan for his son Kim Jong Il to succeed him as supreme leader, he has been attempting to ensure that this dynastic succession will occur.

If propaganda and careful personnel placement were enough to bring his plans to fruition there could be no question that the son would succeed the father. Kim Jong Il is now trumpeted as the next best thing to the god, his father, and efforts are continuing to develop support for him through the appointment to high positions of supporters from his own age group. Yet many questions remain. Dynastic succession is a difficult feat to achieve; successes are few, and it is often the inherent qualities of the successor that count for more than the careful planning of the father. The present leader of Taiwan, Chiang Ching Kuo, inherited power from his father, but there is ample evidence that he was and is a gifted leader in his own right. The same cannot be said of Kim Jong Il, who strikes most observers as a man with few particular abilities of his own. But even if he had such abilities it would not be easy for him to succeed. The dynamics of succession inevitably create enemies, and there are sufficient indications in North Korea's past history to suggest that Kim Jong Il will face opposition, both within his immediate family and among the various power brokers who feel that they should have a role in the process.

It was calculations such as these which gave credibility to reports spreading from South Korea in November 1986 that a coup had been mounted against Kim Il Sung, and that he had been either deposed or killed. For three days loudspeakers along the northern side of the Demilitarized Zone were broadcasting reports of the coup, of Kim's death, and of a seizure of power by the Defence Minister, O Jin U. Yet Kim was televised at the airport meeting a visiting Mongolian delegation on 18 November, just before the last of the broadcasts were heard. Soon afterwards Pyongyang reported that O Jin U had been involved in a serious car accident, which perhaps tended to support the view that he had mounted a coup which had failed.

But few things in North Korea are what they seem on the surface. At the end of December the first National Assembly session since elections on 2 November re-elected Kim Il Sung as President, and it was announced that O Jin U had retained his post as Defence Minister. At a Party Central Committee meeting at about the same time Kim Il Sung, Kim Jong Il, and O Jin U were reconfirmed as members of the Praesidium and Politburo, thus retaining their positions as the top three leaders in North Korea.

What was behind the mysterious broadcasts has not been clarified. It is possible that they were a trap to embarrass South Korea, which hastened to publicize the reports widely and as a result lost some credibility. In support of this thesis is the fact that the broadcasts were limited to only some of the many loudspeakers along the Demilitarized Zone, and none of these could be heard by American listeners. It might also have reflected an abortive effort to gain support for a coup attempt; it is strange, however, that there have not been further repercussions within North Korea.

Even though there is no way to determine if the reports reflected disagreements of a political nature, there have been indications during 1986 that a struggle over economic plans has been under way. In July 1985 Kim Il Sung had announced the formulation of a third seven-year plan. Late in 1985 there was a reorganization of the Administrative Council which made many and significant changes in the economic administration of the country. In February 1986 two Deputy Premiers were replaced, and in August another First Deputy Premier was replaced by the third incumbent within a year. Although the second seven-year plan had been completed in February 1985, and the formulation of the third announced, the economy is now operating in what is officially called a transition period. There are no production targets in such a period, only annual guidelines to follow. Given the poor state of the North Korean economy, the many personnel changes in the upper reaches of the economic machine, and the failure to promulgate the next plan, it is easy to believe that economic problems continue to create political difficulties.

Effect on the Larger Powers

Even in periods of greater stability on the Korean Peninsula than the present one, its history and central position between the super-powers have made it one of the more sensitive areas of the world. The North Korean regime has demonstrated over the years that it is among the more aggressive and unpredictable nations that the international community has to deal with. And, although they have made considerable progress since the end of the Korean War, the South Korean armed forces are still inferior to the North's in numbers and some important types of equipment. Moreover, the close proximity of Seoul to the border creates a special vulnerability to any new strike by the North. In the past two years the North has brought its forces forward (increasing the proportion stationed in the border area from about 45% to about 65%) and increased their ability to move at short notice.

Since the abortive attempt to assassinate President Chun and most of his cabinet in Rangoon in October 1983, North Korea has appeared more conciliatory, opening a series of talks with the South on five different areas of mutual concern which lasted intermittently from late 1984 until January 1986. But since Pyongyang broke off the talks in protest at the annual South Korean–US joint military exercises, only negotiations concerning the forthcoming Olympic Games have been held. It is possible that North Korea prefers to wait and see how the present political difficulties in the South are resolved before committing itself to continuing negotiations; it is also possible that the political undercurrents in the North have contributed to the stand-down. That the North did not seem to have changed its ways, however, was shown on 14 September, just before the opening of the highly successful Asian Games, when a bomb exploded at Seoul's Kimpo airport despite the extraordinary security measures the South Korean government had taken. It was widely believed that this effort to disrupt the serenity of the Games was the work of North Korean agents.

In addition to the threat of normal military action by the North, South Korea has sounded an alarm about the building by Pyongyang of a massive dam and hydro-electric plant on the north Han river. The Kumgang dam, claimed the South Korean Defence Minister, is an uneconomic project being constructed by military labour, which could be used to flood Seoul in advance of an attack from the North. South Korea has said that it will be necessary to take 'self-defensive measures', including building its own dam to hold back the waters from Kumgang, if the North does not cancel its plans.

Both the continued North Korean actions and the South Korean responses illustrate the high level of tension that exists on the peninsula. Should the coming year bring new political uncertainties in both North and South, this tension is bound to be exacerbated. With the Soviet Union increasing its military aid to the North in an effort to increase its influence there, and the United States still the guarantor of South Korea's security, the consequences of a breakdown in the uneasy peace that has existed since 1953 are obvious.

The United States, in particular, has a delicate task before it. Anti-Americanism, fed by student radicals and economic stresses, has grown in South Korea in the past two years. Unless some compromise between President Chun and his determined opponents can be achieved in 1987, tensions are certain to grow on both sides. The US must be careful that President Chun and his military supporters do not mistake its advice to implement liberal reforms for basic disapproval of the government, since that could lead to a strong political reaction in the army against reform. If extremist pressures continue to grow on both sides, 1987 will be one of the most fateful years for Korea since the end of the War.

INSTABILITY IN THE PHILIPPINES

The Philippines remained in an unstable state throughout 1986, with concomitant uncertainty over the future of the large US naval and air force bases there. For the previous twenty years the United States, with its close historic and economic links with the Philippines and with its strategic stake in the country, had looked to the regime of President Ferdinand Marcos to provide stability. But in the last years of his regime, this was precisely what Marcos was increasingly unable to do. In February 1986 he was re-elected in what was widely seen as a heavily rigged presidential election. Two weeks later he flew into exile in the US, after Defence Minister Juan Ponce Enrile and armed forces Deputy Chief-of-Staff Gen. Fidel Ramos had thrown in their lot with Mrs Corazon Aquino, who had been the opposition candidate in the election. Forces loyal to Marcos tried to suppress the coup by an attack on the Enrile and Ramos stronghold in Camp Crame in Manila, but they were prevented by a large crowd of civilians which blocked their way, and on which the Marcos loyalists were unwilling to use force. After this display of what was dubbed 'people power', Mrs Aquino assumed the presidency and declared the need for a 'revolutionary government'. With Aquino's accession to power, the US somewhat hesitantly looked to her government to re-establish stability in the country.

The new president faced three tasks that were clear but inextricably intertwined: to re-establish the democratic political process which Marcos had suspended in 1972 by imposing martial law, and only nominally reinstated when martial law was lifted in January 1981; to formulate an economic policy to stimulate the stagnating Filipino economy; and to solve the growing Communist and Muslim insurgencies in the country. But President Aquino was politically hamstrung from the start by her lack of a strong political base. Although she had broad popular support, she had no political party of her own, and the government she formed from her former allies in opposition was united in opposing Marcos, but in little else. There were still many Marcos supporters in the population, and large sections of the countryside were controlled by Communist and, to a lesser extent, Muslim insurgents. Aquino had only come to power after Enrile and Ramos had switched sides, bringing key elements of the armed forces with them, and it became clear in the course of the year that her position owed much to the support of such elements of the armed forces and their leaders. Her recognition of this was implicit in the appointments of Ramos as Chief-of-Staff of the armed forces and of Enrile as Defence Minister (despite his tenure of the same post under Marcos, and the fact that for the eight years of martial law he had been its administrator).

Mrs Aquino was quick to appoint her cabinet but slow to establish the legitimacy of the new government. In late March, a month after she took power, she proclaimed what she termed a temporary 'freedom constitution', which made the presidency the only constitutional source of legislation by abolishing the national assembly. (It also provided for the establishment of a commission to draft a permanent constitution, to be in place within a year.) To deal with the serious political challenge from local government officials who had been Marcos supporters, some 2,000 'officers-in-charge' (OIC) were appointed to gubernatorial, mayoral and town council posts in their stead. This replacement of elected local officials by OIC appears to have been a serious political error. Many of those ousted might well have been willing to work for the new government if they could retain their local patronage and privileges, and, with their network of connections, would have been a valuable conduit through which to implement its policies. Of the newly appointed OIC, many were clearly opportunists who hoped to acquire the perks their predecessors had enjoyed, while others were idealistic Aquino supporters who lacked administrative or governmental experience.

At the national level, Aquino soon began to appear indecisive, a label which stuck with her throughout most of the year. The cabinet was divided on such issues as trade policy, the tracing and recovery of public funds improperly dispersed by Marcos, and settlement of the insurgency question. In many instances disagreements within the cabinet resulted in slow formulation, and even slower implementation, of policy. The situation was made worse by the impression that the cabinet's efficiency was being undermined by political manoeuvring.

This carried with it two dangers, the first being that it might hinder a solution to the insurgency problem. The Aquino government was trying to reach negotiated settlements with both Communist and Muslim insurgents, and by the end of the year it had declared a cease-fire and opened negotiations with both main groups. But progress in reaching that stage had been too slow for some cabinet members – notably Defence Minister Enrile, who became increasingly outspoken in his criticism of policy towards the insurgents. The second danger was that the government's long-term policies would lack credibility unless it could offer the prospect of long-term cohesion, and this began to seem more and more unlikely.

By the middle of the year rumours of an impending coup had reached a fever pitch in Manila, adding to the general atmosphere of instability. In July, while Aquino was in the south of the country, a 75-year-old maverick elder statesman and former Marcos foreign minister, Arturo Tolentino, launched an ill-conceived coup attempt in Manila, backed by about 200 troops and joined by six generals. It quickly fizzled out once Enrile and Ramos repudiated the rebels, and the episode served to bolster confidence in the commitment of the Armed Forces of the Philippines (AFP) to the President. However, the lack of cohesion inside the cabinet remained. Aquino's distrust of her deputy, Salvador Laurel (who had stepped aside to make way for her in the February presidential elections), was highlighted when she refused to hand over the reins of government to him while she visited Indonesia and Singapore in August and the US in September.

Despite concessions by Aquino on policy towards the insurgents, Enrile's criticism of government policies became more strident. In November, after persistent rumours that soldiers loyal to him and to Marcos were planning a coup, the President overcame her appearance of docility and sacked Enrile. Since the key to this surprise move was the assurances of support she received from Gen. Ramos and other senior officers (it had previously seemed uncertain which way Ramos would jump if forced to choose between Aquino and Enrile), the episode re-emphasized both the AFP's importance in determining who governed the country and the particular importance of Ramos in contemporary Philippine politics. Immediately after this, Ramos and his most senior officers sent Aquino a memorandum which included a 10-point list of proposals on matters such as the conduct of the counterinsurgency campaign and the sacking of certain cabinet ministers. The officers concerned claimed to be acting on behalf of the people, but others interpreted the move as meaning that power-sharing was the price of AFP support. Aquino subsequently dropped three left-wing members of her cabinet in November and December.

The Economy and Insurgency

When the Marcos regime was overthrown it left behind a legacy of economic mismanagement, corruption and profligacy with public funds. The national debt amounted to $25 billion, the debt–service ratio was running at 35% of exports of goods and services, and the GNP had fallen by 3% in the previous year. There had been an explosion in the money supply in the two months before the 1986 election, and Manila was unable to meet targets set by the IMF as a condition for the stand-by credit package agreed in December 1984. In April the Aquino government discontinued that stand-by credit arrangement and by October had succeeded in negotiating a new package worth about $506 million. Negotiations were also opened with the country's commercial bank creditors towards rescheduling debts amounting to $9.6 billion.

Expanding the money supply to stimulate national economic growth was central to the new government's economic strategy. But businesses and investors were slow to react to Aquino's pump-priming measures, primarily because of uncertainty over the country's political stability. By the beginning of 1987 there were some signs that government economic policies had induced a slight increase in manufacturing activity in the previous year, but its political policies had done little to bolster investors' confidence.

Economic recovery – particularly in rural areas – was a key factor in the administration's efforts to deal with the Communist insurgency problem. Aquino appeared to agree with those of her advisers who argued that at least 70% of those in the Communist Party's military wing, the New People's Army (NPA), were not ideologically committed to Communism but had turned to the NPA out of desperation at the economic stagnation and unemployment in rural areas, and had then become tired of life on the run. Most senior AFP officers put the number of committed ideologues in the NPA higher than that.

The present insurgency began in 1969, when the then recently-formed Communist Party of the Philippines (CPP) established the

NPA. In the same year a group of Filipino Muslims formed the Moro National Liberation Front (MNLF) and began a campaign of violence in the south of the country aimed at achieving independence for the Moro people via the establishment of an Islamic state. In 18 years, the NPA has grown from a small, badly-equipped group of fighters into a well-armed and trained force, estimated at more than 22,000, which controls large rural areas spread over some 60 of the country's 73 provinces. Starting in the mid-1970s, the CPP has moved away from its Maoist image and has been developing a nationalist Communist ideology. There have been reports that the NPA is now receiving aid from Moscow, but there is no substantial evidence to support this.

In addition to addressing the economic aspects of the insurgency problem, the Aquino government worked towards a cease-fire and negotiations with both the Communists and the Muslims. In March it released detained Communist leaders along with other political detainees imprisoned by the Marcos regime, and in August formal cease-fire talks with Communist representatives began. While they were in progress, however, the AFP suffered heavy losses, and Enrile and senior AFP officers put the government under heavy pressure to achieve results. Army units were ordered to resume offensive activities against the insurgents in September, and in November an agreement was reached for a 60-day cease-fire to begin on 10 December.

All sides used the pause in the fighting to fight a propaganda war. The day the cease-fire began the government announced a programme of national reconciliation and development aimed at revitalizing economic areas by creating self-sufficient communities. The plan was tied to a scheme to resettle NPA fighters who surrendered. Negotiations then began between the government and the National Democratic Front (NDF), a CPP-led umbrella organization, for a more lasting peace settlement. The NDF was believed to be asking for radical land reform, the restructuring of the AFP, and the removal of US military bases from the country. It is unlikely that such demands would have been acceptable to AFP leaders.

The Future of the US Bases

Although political negotiations are probably no threat to the US bases in the Philippines, the possibility that Communist insurgents might eventually win power in the country poses an important long-term threat to the US military presence there. The NPA's fighters are operating in the provinces around the Clark air and Subic Bay naval bases, but they have apparently been careful not to attack US military facilities or personnel for fear of precipitating greater American involvement in Manila's war against them. Nonetheless, if the Communists ever do take power, they have said that they are committed to the removal of the US bases. This gives the United States a powerful incentive to offer any assistance, economic or military, that Manila sees as necessary to defeat the insurgency.

Throughout 1986 the attitude of the Aquino government towards the US bases reflected the traditional Filipino problem of reconciling nationalism with the economic and security needs which the US presence fills. Although she had signed a declaration a few years ago calling for the removal of US bases from the Philippines, Aquino had modified her position even before she became president. When she came to power, she said that she would attempt to renegotiate the bases agreement and then hold a referendum asking the people of the Philippines whether they wanted the US bases to remain. But she gave no timetable for this, thus leaving her options open about what would happen once the bases agreement expired in 1991.

The major military bases and facilities that the US has in the Philippines were established after Manila and Washington signed a 99-year Military Bases Agreement (MBA) in March 1947. The agreement has since undergone major revisions, and in its current version is subject to five-yearly reviews (the next will be in June 1988) and is due to expire in 1991. Each review is accompanied by a financial agreement (referred to as 'aid' by the US and 'rent' by the Philippines); at the June 1983 review President Reagan agreed to try to procure $900 million in defence and economic assistance over five years beginning in October 1984. The terms under which any renewal of the MBA will be negotiated after 1991 will be dictated at least in part by Article XVIII, Section 25 of the new Philippines Constitution, which stipulates that after the expiry of the MBA foreign troops and military bases will only be permitted in the Philippines under a treaty ratified by the Philippine Senate and, if Congress requests it, subject to majority approval in a national referendum.

The two main bases include Subic Bay naval base (home of the US 7th Fleet), which is used for about two-thirds of the fleet's support functions, and Clark air force base (home of the US 13th Air Force), which has a 10,500-ft all-weather runway capable of taking the largest aircraft in the USAF inventory. The two bases have many advantages over alternative locations in the Pacific. For instance, they are close to the Sunda and Malacca straits, through which some 5 million barrels of oil pass every day, including 65% of Japan's supply. In addition, they offset the increase in the Soviet presence at the Vietnamese bases of Da Nang and Cam Ranh Bay, only about 800 miles to the west. For the ASEAN states, such a counterweight to Soviet military facilities in Vietnam is important for regional security. The USSR is reported to have facilities at Cam Ranh Bay for surface warships and submarines, while at Da Nang there are reconnaissance and bomber aircraft, including Tu-16 *Badgers*, MiG-23 *Flogger* fighters and mid-air refuelling aircraft, and in addition an electronic monitoring system is maintained there.

Contingency planning has been going on in Washington for several years against the possibility that the US might decide or be forced to withdraw from its military bases in the Philippines. Recent events have sharpened American concern over the future of the bases, but it would be both difficult and expensive to find new homes for the forces involved. In June 1983 Admiral Robert Long – then Commander-in-Chief, Pacific – told a Congressional Committee that to rebuild the facilities at Clark and Subic on alternative sites in Guam, Palau and the Marianas would cost a minimum of $3–4 billion in fixed investment alone, and would take five to six years. And a report commissioned by the US Senate Appropriations Committee, submitted in March 1986, estimated that it would cost up to $8 billion to move US military facilities from the Philippines. This report envisaged the dispersal of US forces in the Philippines to existing air force and naval facilities in Guam, Okinawa, Saipan and the Tinian islands in the Marianas, and perhaps to US bases in South Korea and Japan. Operational costs at these facilities would be much higher, since the absence of a trained local workforce would mean using US personnel, instead of a Filipino workforce, at much higher rates. Another problem in relocating the bases is the unwillingness of other possible hosts to accept them: there is strong local opposition to a major base on Palau, and Guam already accommodates so many military facilities that there would be a further price to pay for relocating there, in terms of winning local support.

Perhaps the most expensive consequence of moving from Clark and Subic would be the need to man and build more ships and aircraft in order to sustain the present US strength in the western Pacific from bases in the central Pacific. Strategically, too, such an arrangement would be disadvantageous. The complementary nature of the two main bases would be lost, logistic support would be dispersed and much of it relocated thousands of miles further away from such potential theatres of operations as the South China Sea and the Indian Ocean, complicating battle planning and slowing response time.

In December 1986 the commander of the US Pacific Fleet, Admiral James Lyons said the bases in the Philippines were key to the US remaining a western Pacific power. Although this is the prevalent attitude in Washington, there is some debate as to whether these bases are in fact vital to the US. This reappraisal appears to have been prompted by a change in the focus of US strategic priorities in the Pacific from containing Communist China to balancing the Soviet military build-up in the Far East and Pacific. In addition some analysts argue that the US no longer needs a large logistic base in the western Pacific because of the increased ranges of naval ships, the use of under-way replenishment ships, the development of mid-air refuelling, the increased range of carrier-borne aircraft, and the use of long-range missiles. While such arguments are unlikely to be acceptable to Washington at present, their proponents may grow in number. If they do, it may become increasingly acceptable in Washington to contemplate relocating US military facilities in the Philippines to less satisfactory sites, provided the US economy can absorb the high costs involved.

SECURITY PROBLEMS IN THE SOUTH PACIFIC

The security treaty between Australia, New Zealand and the United States (ANZUS), signed in 1951, has now ceased to exist as a trilateral defence agreement. David Lange's New Zealand Labour government tried hard to persuade Washington to keep the treaty's main elements in force, despite New Zealand's ban on port visits by US and British naval vessels with nuclear arms or propulsion. But this was a circle it was unable to square. The August 1986 communiqué of the Australian–US ministerial talks (replacing the ANZUS Council) declared unequivocally: 'Both governments regretted that the continuation of New Zealand's port and air access policies has caused the disruption of the alliance relationship between the US and New Zealand. They agreed that access for allied ships and aircraft is essential to the effectiveness of the ANZUS alliance, and that New Zealand's current policies detract from individual and collective capacity to resist armed attack'.

In formal terms, the treaty remains in place; in practical terms, all trilateral and US–New Zealand activities are, at the very least, suspended. Australia retains all its links with both the US and New Zealand, but on separate bilateral bases. Meanwhile, New Zealand – which cannot see why it is not treated like Denmark – hopes the United States will change its policy over membership responsibilities; the US and Australia hope for a change in New Zealand's policy on nuclear ship visits. Neither hope is likely to be fulfilled in the near future. Even a Democratic US President would find it hard to reverse policy, with all the accompanying publicity and the implications for the Western alliance system.

New Zealand, as one of its commentators said, is deeply divided and deeply confused. Tests of public opinion by a Defence Committee of Inquiry which reported in August showed that 72% of the community favour an alliance with larger countries (US, Australia); but 73%, many of them the same people, want their defence to be managed in a way which ensures that New Zealand is nuclear-free. In a different framework, 52% of respondents favoured a return to an operational ANZUS alliance and the acceptance of visits by ships which may or may not be nuclear-armed or nuclear-powered, but 44% would prefer to withdraw from ANZUS rather than accept nuclear ship visits.

While the tone of the Defence Committee of Inquiry's report ran counter to New Zealand Labour government philosophy and prompted some tart exchanges with the Prime Minister, its recommendations were acceptable. These included an enhanced bilateral relationship with Australia, greater use of scientific and technological expertise, professional armed forces with a greater degree of self-reliance, and an enlarged intelligence-gathering and assessment capability. There is no evidence yet that the government will allocate to defence the additional funding it needs, and the legislation preventing air and ship visits is making a very leisurely progress through Parliament. In the process of carrying out a defence review, the government decided in December 1986 to follow the earlier examples of Britain and Australia and withdraw its small infantry battalion and air force contingent (which totals 740 men) from Singapore by the end of 1989, marking the virtual end of more than thirty years of active defence co-operation. The Five-Power Defence Arrangements are still in force, and an Australian fighter squadron is still stationed at Butterworth in Malaysia, but this new withdrawal must make them more difficult to implement, and thus less credible and less relevant.

For its part, and despite the fact that New Zealand's defection in some ways strengthened the US–Australian alliance relationship, Australia found its super-power ally a more difficult partner than before. The problem was not nuclear policy, but the competition created for Australia's wheat and sugar in export markets by subsidized American exports of the same commodities. This produced strong reactions from the Australian government, which for the first time called the alliance into question with Foreign Minister Hayden asking what kind of a partner would subvert its ally's economy. The US, which had termed New Zealand a friend but not an ally, was now in effect calling Australia an ally but not a friend, and Prime Minister Hawke accused it of abandoning its principles for short-term political expediency. Australian complaints had no effect on American policy, however, and the suggestion that the Australian government might threaten the leases on US facilities in Australia as a countermeasure was not taken seriously by either side.

Like New Zealand, Australia has been reappraising its security. In 1985 the Minister for Defence, Kim Beazley, commissioned Paul Dibb of the Australian National University (and formerly Deputy Director of the Joint Intelligence Organization) to examine 'the extent, priorities and rationale of defence forward planning in the light of the strategic and financial planning guidance endorsed by the Government' and to advise on which defence capabilities should most appropriately be developed. The Dibb Report, well researched and well argued, was presented to Parliament in June 1986 and was accepted by the Cabinet. It rationalized something that had been taking place over the previous decade: a change of emphasis in Australian defence thinking away from a strategy of defending the country as far from its shores as possible (loosely known as 'forward defence') towards a 'strategy of denial' on land and in the sea and air environment immediately around it.

Dibb considered that Australia was one of the most secure countries in the world, that it faced no direct, identifiable military threat, and that there was every prospect that this favourable circumstance would continue. He accepted official strategic guidance that it would receive at least ten years' warning of a substantial military threat – a convenient thesis, if historically improbable – while acknowledging that low-level contingencies were much more credible. Because of the country's vast size and scattered settlement, responding to such contingencies would require a disproportionate, but fundamentally local response. He thus placed his emphasis on the mobility of land forces and on naval and air forces capable of denying an adversary control of the sea and air gap between Australia and nearby foreign territory. Although aware of the problems posed by shipping choke points, Dibb concluded, somewhat contradictorily, that 'most military activities involving disruption of Australian trade could be handled by evasive routing'.

But the key element in his report (and the crucial point of difference from virtually all previous Australian defence strategy this century) lies in the assumption that Australia is, or ought to be, basically a consumer and not a producer of security in its region. Dibb agrees that ANZUS is beneficial in all sorts of ways – intelligence information, logistic support, weapons acquisition, and defence science and technology – and that US facilities in Australia contribute to Western deterrent capacity; but he takes the view that *'there is no requirement for Australia to become involved in United States contingency planning for global war'* (emphasis in the original). Similarly, while a future government might wish to have the option of making a modest military contribution in support of 'our more distant diplomatic interests and the military efforts of others', this should be seen as a gesture, and Australian military forces should not be specifically structured or equipped to undertake such tasks.

Although not anti-American in tone, the report distanced itself from some of the more controversial policies of the Reagan Administration in the same way that Canberra had refused to co-operate with the US Strategic Defense Initiative research activities. While criticized by some for being too defensive in orientation, the Dibb report was generally well received by government and the public. It fitted the philosophy of many in the ruling Labor Party, and the logic of its arguments satisfied many conservative elements in the country.

The Wide Ocean Spaces

There has been low-level but continuing turbulence in recent years in the vast area of Oceania. Most of the small island states are economically weak and politically inexperienced, and the turbulence is due

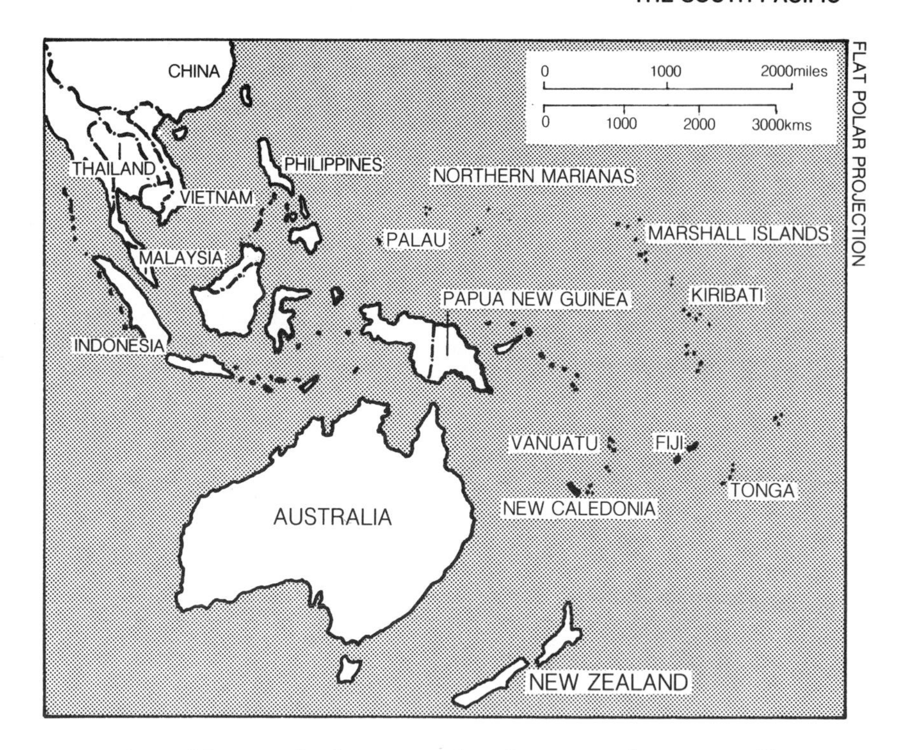

as much to this as to the interests of major external powers – France, the United States and the Soviet Union.

The island states' perennial displeasure with France, over its nuclear testing programme in French Polynesia and its colonial presence in the Pacific, was increased in 1986 by the change in policy towards New Caledonia effected by M. Chirac's new conservative government. The population of New Caledonia is 42.5% Melanesian, 37.1% European, 12.2% Wallisian and Polynesian, and the remainder a mix of Indonesian, Vietnamese and others; not all Melanesians want independence, nor do all Europeans want to remain tied to France. Elections held in September 1985 had produced indigenous Melanesian majorities in three of the four local government regions, while whites controlled the island as a whole. President Mitterrand and his Socialist government had assured the Kanaks (Melanesians) of their right to choose independence, preferably 'in association with France'. However, the Chirac government, elected in March 1986, quickly rejected this and reduced the power of local governments under interim arrangements until a referendum could be held in the summer of 1987. The electoral arrangements seem likely to produce a majority for the *status quo*, and the coalition of seven indigenous groups known as FLNKS (Kanak Socialist National Liberation Front) seems certain to boycott the referendum.

The independence movement in New Caledonia has been encouraged by the South Pacific Forum – a thirteen-member body made up of the states of the region and including Australia and New Zealand. At its annual meeting in August the Forum, angered by Chirac's policy change, unanimously agreed to propose to the UN General Assembly that New Caledonia should be classified as a non-self-governing territory subject to international control, and should no longer be treated as a French overseas territory. Support for this move from Australia (somewhat reluctantly) and New Zealand drew the wrath of M. Chirac, who described the Australian Prime Minister as 'very stupid' and warmly praised the two French agents who had blown up the Greenpeace vessel *Rainbow Warrior* in Auckland harbour in July 1985, killing one of the crew. (New Zealand, effectively capitulating to French economic pressure, had released these agents into French custody in the summer of 1986 after negotiations by UN Secretary General Pérez de Cuellar). On 2 December the UN General Assembly voted to add New Caledonia to the list of territories that should be independent and asked France to co-operate. In January 1987, as a further indication of French displeasure, the Australian Consul-General in New Caledonia, Mr John Dauth, was declared *persona non grata* without explanation, and France suspended ministerial contacts with Australia.

France is unlikely to budge from its Polynesian territory until it no longer needs the Mururoa nuclear testing site. In the meantime, it has difficulty in understanding the Australian attitude. The French nuclear capacity, it believes, contributes to Western security by complicating Soviet calculations of the consequences of even a conventional attack on Europe's defence, so that the USSR is less able to use the threat of nuclear arms to intimidate other states. To the Pacific states, these arguments would be more convincing if France did its nuclear testing at home. The US finds the reasoning acceptable, but then the US has its own missile testing programme in the Pacific.

It is because of that programme that the US (like France, but unlike China and the Soviet Union) has refused to sign any of the protocols to the Treaty of Rarotonga. These establish a nuclear-weapons-free zone in the region, banning nuclear-weapons states from stationing, testing, using or threatening the use of nuclear weapons or devices there. Of the thirteen South Pacific Forum member states, ten have signed the treaty, and enough (eight) have ratified it to bring the treaty into effect, limited though its provisions are. The Soviet Union has gained some kudos from its signature, which is seen by Western states as in accord with the gentle but persistent Soviet diplomatic effort in the region set out in Mr Gorbachev's Vladivostok speech in July.

In 1985 the USSR signed a one-year fishing agreement with Kiribati, paying a fee to fish in that state's exclusive economic zone; the agreement was not renewed on expiry, although Kiribati later said it would welcome a renewed approach. In the meantime Moscow has concluded a similar but more comprehensive treaty with Vanuatu and has apparently made approaches to Fiji, Tonga and Papua New Guinea, perhaps intending to play one off against another. Flattering personal diplomacy is now official Soviet policy in the region: its Ambassador to New Zealand, for example, even went to church during his visit to Tonga, and added 16 balalaikas to the King of Tonga's collection of musical instruments. One of the Pacific states' motives in talking to the Soviet Union has been to exert pressure on the Reagan Administration to compel American tuna boats to respect their exclusive economic zones. The Soviet (and Cuban) overtures in the region eventually attracted Washington's attention, and in October an agreement was signed with the South Pacific Forum Fisheries Agency which granted the US non-exclusive fishing rights for five years, at a total cost to the US of at least $60 million. That Japan, too, had new concerns in the region was shown in January 1987 when Foreign Minister Kuranari visited Australia, New Zealand and Fiji; in a major speech in Fiji he spelled out Japan's diplomatic and economic aid interests in the area.

The contacts that have developed between indigenous independence movements and the Soviet Union have jolted the US into reviewing the status of the US Trust Territory of the Pacific. After appropriate legislation and referenda in the component states, a US Presidential Proclamation on 3 November declared the Trusteeship Agreement no longer in effect. A 'compact of free association' had already been signed by the United States (severally) with Palau, the Marshall Islands and the Federated States of Micronesia, giving the US control over foreign and defence policies, while the Northern Marianas opted for Commonwealth status with the US (equivalent to that of Puerto Rico). Although accepted by the UN Trusteeship Council, and theoretically offering greater stability to the US presence, these arrangements (with their accompanying US subsidies) will not necessarily resolve the legal or the political situations of the territories nor contain the indigenous pressures for independence. A court in Palau has ruled that the Free Association compact violated the nuclear exclusion clauses of the Palauan constitution and had not been adopted by a majority sufficient to override the constitution. Therefore, unless its constitution is amended, or some other compromise is reached, Palau will continue as a Trust Territory, and the US will retain its rights there as the administering power; but the majority that voted for self-government will feel they have been denied it.

Africa

LIVING WITH CRISIS

If the first half of the 1980s was a period of deep recession for the industrialized world, the second half is proving to be one of prolonged depression for Africa. Despite its wealth of natural resources, the region has been singled out as the riskiest in the world for investment. In 1986, agricultural output grew for the first time in fifteen years, increasing by about 3%, or slightly more than the population growth rate. But the continent's external debt grew also. Now estimated at about $200 billion (up from $150 billion in the previous year), it amounted to a staggering 44% of the continent's GDP and some 190% of its export earnings. Although imports during 1986 fell by nearly 21%, the overall balance-of-payments deficit for the year was expected to reach $25 billion, more than twice the level for 1985. Commodity prices, which dropped with the rise in the dollar, have not so far responded favourably to its subsequent fall; export earnings thus fell from $60.6 billion in 1985 to $44.3 billion in 1986 (a drop of some 27% overall, but for major oil producers, such as Nigeria, Libya and Algeria, revenues dropped by as much as 42–47%). The debt crisis is as serious today as it has ever been, and there is little relief in sight.

Everyone agrees that African states caught up in this cycle of impoverishment must embark on reforms to secure self-sufficiency, attract foreign investment and provide incentives. But on the political, as well as the economic, level much of Africa is paralysed by crisis or corruption and governed by regimes which are the product of military coups or are maintained in office as a front for military power-brokers. With only a few gratifying exceptions, the leadership necessary to cope with the area's deep economic and social problems is generally lacking, and social unrest, guerrilla insurgency and civil war have been the all-but-inevitable results. Nature itself has added its measure of menace: while the devastation threatened by locusts was averted in 1986, the UN Food and Agriculture Organization warns that another locust plague now on its way is likely to destroy the majority of sub-Saharan crops in 1987 unless adequate precautionary measures are soon taken.

Nigeria: The Path of Reconstruction

Notwithstanding the general atmosphere of economic decay and political stalemate across the continent, countries vary from one to another in their ability and readiness to confront the problems before them. In an area of the world where violent change of government is almost an annual event, and where military governments usually fail to live up to the promises made on seizing power, President Ibrahim Babangida's military regime in Nigeria has been unusual in bringing back a feeling of security and putting the nation's recovery back on track. His decision to set a target date for the return to civilian rule – 1 October 1990 – has promised a new political future for the nation and may satisfy popular aspirations for a substantive change. The ten-year ban on politicians of the second republic has been generally welcomed, but is likely to generate considerable debate as well.

An ailing economy, made more vulnerable by the unexpectedly sharp drop in oil prices, has prompted tough economic reforms including the creation of a free foreign-exchange market, known as the Second Tier Foreign Exchange Market. After the government rejected IMF proposals for structural adjustment the national currency lost 50% of its foreign-exchange value, and observers are divided over whether the regime is to be congratulated for avoiding the IMF remedies or condemned for urging further sacrifices on its people. However, agreement was reached with a group of London banks to reschedule medium-term debt repayments, and the World Bank has agreed to provide a $4.8-billion loan. The political climate in Nigeria, unlike that in so many other countries, looks healthy at the moment, and a return of higher oil prices may help the government to lead the country to a new political consensus upon which civilian government, if it resumes in 1990, can build for the future.

Chad: Libya on the Run

Throughout 1986 fighting continued in Chad, but in the late autumn and in early 1987 the nature of the war changed dramatically. Goukouni Oueddei – leader of the *Gouvernement d'Union Nationale de Transition* (GUNT), whose troops had been seeking to overthrow Chad's President Hissène Habré – broke with Col. Gaddafi of Libya, who had been supporting him in the north of the country. Acheik Ibn Oumar was subsequently elected as the new GUNT leader and is now the only Chadian figure directly linked with Libya; his small following numbers about 1,000 supporters.

Virtually all Goukouni's troops in Chad pledged support for Habré: some of them defecting south to join Habré in the capital, Ndjamena, others remaining in the north to help to expel the Libyan troops with whom they had previously been allied. Owing to this diplomatic revolution, which turned the civil war into a war to defeat the Libyan occupation, Habré was able to prepare plans to reconquer the whole of Chad, and on New Year's Day 1987 he inflicted a crushing defeat upon Libyan forces which had been occupying the important northern towns of Fada and Zouar. To help in dealing with the Libyan air threat France chose to reinforce Habré in the south, sending further aircraft and troops to Ndjamena, and also redeploying some forces to Abéché, near the border with Sudan. The

United States contributed $15 million of military aid in late 1986 and also helped to transport some of the French reinforcements in *Galaxy* aircraft stationed in Europe.

The increased French support in 1987 accomplished two things. First, it helped to seal the Chad–Sudan border – Habré and his French advisers had been convinced for some time that Gaddafi would try to open a new front by sending elements of his Islamic Pan-African Legion through Sudan, and the increased French presence in eastern Chad helped to guard against this. Second, it meant that Habré could devote all his resources to the war in the north, taking advantage of new equipment (especially *Milan* anti-tank missiles) that France had provided. In March 1987 he gained a spectacular victory, taking control of the large airfield at Ouadi Doum and so depriving Libya of its principal air base in Chad. This put him in an excellent position to retake the large oasis of Faya Largeau (his birthplace) from which demoralized Libyan troops were streaming northwards.

The question that Habré's successes pose is whether the new alliance he has formed with his old enemy Goukouni Oueddei would survive a victory over Libyan forces. As the battles raged in the first months of 1987 Habré's representatives were negotiating with Goukouni in Algeria. For the first time Habré has recognized Goukouni as a leader with whom he must deal. But in Chad war is a form of politics by other means, and if peace does come to the country, it is not certain that factions temporarily allied to fight for their country would not once again divide against themselves.

Uganda: Museveni Begins His Second Year

The record of President Yoweri Museveni's first year in office provides some hope that he may bring to an end the factionalism and tribal conflict that have plagued Uganda ever since its independence from Britain in 1962. As the victor in a bitter guerrilla war, he might have been expected to impose the usual retributive settlement that accompanies a 'revolutionary' regime; instead he has demonstrated restraint and a preference for civilian legality, even towards his defeated opponents. Reconstruction here requires even more than the usual improvements in administration or elimination of incompetence and corruption. The state had become the enemy of the people: terror and violent death had been rendered commonplace. Museveni therefore has the formidable task of restoring to the population a sense of security and the protection of law.

In this he faces considerable obstacles. Forces loyal to former presidents Milton Obote and Tito Okello still roam areas in the north, despite the regime's assurances that 'troublemakers' have been 'neutralized'. Most difficult of all are the economic problems arising from the breakdown of the country's infrastructure: unacceptable inflation and a foreign debt of some $1.5 billion which entails an annual servicing bill of some $200 million. His unorthodox policies have so far failed to attract foreign investment, so that he must find some other means – barter arrangements, debt cancellation or rescheduling of payments – to break out of the economic chaos inherited from his predecessors. Much of his economic credibility will depend on his ability to overcome tribal disunity and political instability in the north; and yet much of his ability to do so will ultimately require some sign of progress in getting the economy back on an even keel. Museveni's second year in power may well prove to be an even more demanding test of his leadership than the first.

Sudan: The Civil War Continues

When Sadeq el-Mahdi became prime minister in May 1986, after the first democratic elections since Gen. Nimeiry had seized power in 1969, hopes were raised that the event might signal the end of the protracted civil war being waged in the south of the country by Col. John Garang's Sudanese People's Liberation Army (SPLA). In the event, the political will to achieve a solution was insufficient: in fact, fighting resumed with unprecedented violence. Famine became a deadly new weapon in the arsenals of both sides, as the government and the SPLA struggled to control the relief supplies intended to save some two million people from the threat of imminent starvation. Khartoum tried to prevent relief supplies reaching SPLA-controlled areas; the SPLA, on the other hand, claimed that the supplies were being channelled to government forces and shot down a Sudan Airways aircraft which flew over territory it controlled. A crisis point was reached in August/September, but food supplies eventually broke through the blockade, and further appalling consequences of the famine were averted.

El-Mahdi's initial approach to the SPLA hardened as he found himself confronted by multiplying problems of economic dislocation, social unrest and unreliable military forces. Fierce rioting in Khartoum, reminiscent of that which preceded the downfall of Nimeiry, was triggered by a student protest march which the police suppressed. Despite its isolation in the assembly, the Islamic National Front (INF), a fundamentalist movement with student support, can still pose a threat to el-Mahdi's coalition government by resorting to what its leader, Hassan Turabi, calls 'the power of the streets'. The INF looks to the army as the ultimate salvation of the country and the real arbiter of its political choices. Recognizing this – and conscious both of the poor performance of government forces and the vulnerability of his air force to SPLA surface-to-air missiles – el-Mahdi has undertaken an overhaul of the armed forces and toured countries as diverse as the US, Libya, the USSR, Britain and Egypt in search of the arms and assistance he needs to carry it through.

The daily cost of the war in the south is put at $1 million, already an exorbitant expense for a country with Sudan's economic problems, and reorganizing the armed forces may well impose a further massive drain on resources which the country can ill afford. With its total external debt of some $12 billion, and desperately needing further loans as well as a moratorium on interest payments, Sudan has made it clear that it will limit its repayment to 25–30% of expected exports in 1986–7.

Some resolution of the SPLA insurgency is clearly vital to the continuity of order and any hope of recovery. The SPLA, whose strength is in rural areas, insists on the abandonment of the religious fundamentalist constitution imposed by Nimeiry and a return to the previous secular constitution. In April SPLA representatives met a delegation of trade unionists, politicians and professional leaders in Addis Ababa to produce the Koka Dam declaration, a six-point agenda for national dialogue. El-Mahdi has indicated some readiness to meet opposition demands for repealing *Sharia* (religious law), but to agree to the abrogation of treaties with Egypt and Libya could irritate powerful neighbours just when he needs all the foreign support he can muster. If, as it seems, the SPLA is fighting for autonomy, rather than secession, there may be some middle ground for concessions and agreement if both sides are prepared to put national interests above partisan considerations. In the last – and tragic – analysis, it may well be that the spectre of famine will ultimately dictate the terms on which internal struggle is settled.

The Maghreb: A House Divided Against Itself

Col. Gaddafi's whimsical ways, which have so often perplexed the great powers, have also imposed an unpredictable tempo on the dance of diplomatic ambitions in the Maghreb. Once allied with Algeria, and a fellow supporter of the Polisario guerrillas fighting King Hassan of Morocco, Libya changed partners in August 1984 and signed the Oujda Treaty with Morocco – a development which apparently took Algeria's President Benjedid by surprise. As Libyan arms supplies to Polisario ceased (although some funding was said to continue), Rabat bolstered its defences along the frontiers with Algeria and Mauritania. But the King's July 1986 meeting with Israel's Shimon Peres, in an attempt to restart the Middle East peace process, prompted predictable criticism from Gaddafi; Morocco replied by revoking the treaty of Oujda, which by then had doubtless served its purpose so far as Hassan was concerned. This breach has pushed Gaddafi once more into friendship with Algeria's Chadli Benjedid, who visited Libya twice in one month to hasten its return to the camp of the Arab hard-liners.

While 1986 was economically the worst year in a long time for most of the Maghreb, for Morocco it proved to be the most favourable for a decade. To be sure, Polisario claims to the Western Sahara, for which Algeria provides support and a haven for guerrilla fighters, remain an unresolved problem. Hassan has declared his willingness to abide by a unilateral cease-fire and accept a UN-sponsored plebiscite, but so far hopes of such a tidy resolution have proved illusory. Polisario has managed to acquire diplomatic recognition from 66 countries, but cannot count any Western industrialized states or members of the Eastern bloc among them.

In the past year Polisario's difficulties have been greatly increased by Morocco's success in sealing its borders against incursion. The 'Hassan Wall' – earth and stone fortifications six metres high, radar monitored to permit rapid deployment of defensive forces – was regarded with some scepticism initially, but has proved a formidable military defensive system: two unsuccessful attempts to breach it in August resulted in heavy casualties for Polisario's guerrillas. (They did have greater success in March 1987, but not enough to suggest that the wall has done anything other than change the course of the war completely.) Morocco's defensive tactics enable it to stalemate the Polisario offensive, but their very effectiveness may heighten the risk of involvement by Algerian forces, no longer able to operate through their guerrilla proxies. This in turn increases the possibility of direct confrontation between the two countries – something which King Hassan, with the example of the Gulf war before him, is anxious to avoid.

For its part, Algeria, along with Libya (and, to a lesser extent, Tunisia), is suffering from the drop in oil revenues. With over $18 billion in foreign indebtedness, Algeria is the largest single debtor in Africa; barring a spectacular recovery of oil prices to provide relief, it will probably seek rescheduling of its debt in 1987. Social unrest has been the inevitable side-effect of economic strain. In November, riots in Constantine (which soon spread to several other towns) were put down firmly, but the underlying problems remain. Efforts to promote a limited liberalization of the economy have met with stiff opposition from socialist hard-liners in the FLN, Algeria's only political party.

Tunisia has adopted the stance of interested bystander in the contention of Maghreb politics and sought a middle ground between Algerian and Moroccan rivalry. But it faces a period of continuing political vacuum and uncertainty because of the absence of a clear successor to its ailing president, Habib Bourgiba. (Mohamed Mzali, who was Prime Minister briefly in June, was replaced after a month by Rachid Sfar.) It is possible that Bourgiba's exiled former wife, Wasila, might return to take up the task. Without some clarification of its future leadership, Tunisian politics constitute an unknown quantity in an area of conflicting interests and sudden realignments.

Libya's future, of course, is another unknown factor, but for different reasons. While it seems to have weathered the US air strike in April, Gaddafi's regime continues to feel the pressure of American diplomatic and economic hostility, and has been reduced to appealing directly to the population to sacrifice a month's salary to pay for arms purchases – which, apparently, petrodollar earnings are insufficient to cover. Gaddafi has also launched a radical proposal to set up a million-strong volunteer army to be run on the lines of his Revolutionary Committees, arguing that the standing army is expensive and contrary to his ideals of participatory democracy. Most analysts, however, believe that he sees the professional soldiers – irritated by his strategy in Chad and the wasting on foreign adventures of funds which could be better spent at home – as the main threat to his authority, and that he hopes to keep them in line by means of a parallel force of volunteer youths totally loyal to his leadership. So far, his efforts to outmanoeuvre Morocco's King Hassan and undermine Egyptian President Mubarak have backfired, and his attempts to play puppet-master in Chad and Sudan seem as accident-prone as the enthusiasm he has shown for merging his state with others over the years. A revival of oil prices could yet restore much of his influence in the region and his authority at home, but continued experimenting with the explosive chemicals of regional and international politics could also lead to further problems for his leadership.

In 1986, the military regime in Mauritania initiated the first municipal elections in the country's 26-year history. Further elections planned over the next two years create hopes that a process has begun which could lead to the establishment of a multi-party system. Such progress has been made despite violent protests staged in the capital, Nouakchott, by the clandestine African Liberation Forces of Mauritania (FLAM). But the economy remains gravely ill, and there appears to be no prospect of a speedy recovery.

In general terms, the region is still plagued by economic problems and political ambitions which hopelessly divide its member states and all but rule out any reasonable prospect of a united Maghreb. Libya, Algeria and Morocco are still stockpiling sophisticated arms while they grapple with problems of poverty and political uncertainty. Although the region's leaders have become more restrained in their criticism of each other of late, there seems to be little evidence of any policy of co-operation and compromise developing that might lead to a consensus among them on how to deal with the difficulties they face in common.

SOUTHERN AFRICA: CONFLICT AND DISRUPTION

Southern Africa's several little wars spilled across more borders in 1986, for the first time affecting all ten regional states and causing economic and social disruption on a scale likely to outlast the conflicts themselves. Indeed, the ability of the Angolan and Mozambican governments to continue functioning in the face of staggering economic and military pressures raises questions about the validity of the notion of economic collapse of third-world governments.

In Mozambique the death of President Machel in October threatened to weaken further a government already plagued by massive starvation, economic paralysis and inability to make headway against the Mozambican National Resistance (RENAMO): an insurgent group whose lack of credentials as a national movement has not prevented it from waging an effective guerrilla war against the government in every one of the country's provinces. By early 1987, however, the new Chissano government, buoyed by increased support from the Western powers and the Front Line states, seemed to have weathered the initial transition period without serious splits or frictions among the leadership. President Chissano pledged to follow the pragmatic policies of his predecessor, sought closer links to the West, and publicly rejected the idea of a compromise with RENAMO.

The conflict in Angola reached at least a temporary stalemate in 1986, as neither the government's increased offensive capability nor the guerrillas' acquisition of US *Stinger* surface-to-air missiles significantly altered the relative balance of forces. The UNITA insurgents continued to carry out hit-and-run attacks across large areas of the country, and to lay mines to deter peasants from tilling their fields. But UNITA remains unable to hold or control any territory beyond its stronghold in Angola's remote south-eastern corner and a salient on the south-central plateau which straddles the vital Benguela railway. With direct support from South African forces, UNITA was able to blunt the government's latest offensive before the rainy season brought a temporary halt to the war, but there is a strong likelihood that the government will renew its assault on UNITA's strongholds in 1987. The escalation of conventional warfare, possibly involving armed clashes between Cuban and South African forces inside Angola, remains a real possibility.

Diplomatic and military activity in Namibia continued at a low level during 1986. Prospects for new Western peace initiatives, already stalled by long-standing divisions between the US and its European allies over Namibia policy, were further dampened by the growing tensions between South Africa and the industrial democracies over Pretoria's belligerence towards neighbouring states and domestic dissidents. The hard-hitting counterinsurgency operations of the South African Defence Forces (SADF), together with harsh police measures inside Namibia, continued to keep the South-West Africa People's Organization guerrillas from posing a serious military threat. Despite heavy casualties incurred over the past few years, however, SWAPO was able to field a force of 1,500 guerrillas (by a South African count) for infiltration into Namibia in 1986. Also in 1986 SWAPO's first legally sanctioned political rally inside Namibia drew a turnout of 10,000: one of the largest yet seen in Windhoek. Meanwhile Namibia's transitional government, representing a group of moderate ethnic parties (and excluding SWAPO), met with President Botha to seek greater autonomy for the Territory.

All these regional developments were overshadowed, however, by the rising spiral of protest, repression and violence inside South Africa. As the industrial democracies imposed economic sanctions late in the year, the Botha government introduced temporary punitive economic measures of its own against several neighbours, thus highlighting their economic vulnerability and serving notice that sanctions would not be allowed to affect South Africa alone. Internally, the government failed to put a complete end to the racial protest and disruption in its black communities despite the imposition of the most sweeping security and censorship measures outside the Communist world. In addition, 1986 was the fourth successive year of deep economic recession, with high rates of unemployment spreading to the white community for the first time in forty years and foreign bankers and investors showing little confidence in the government's will or ability to deal effectively with the crisis.

After months of indecision and temporizing in Pretoria, the Botha strategy for dealing with the crisis emerged late in the year. Characteristically, Botha opted for an uncompromising line towards both domestic and foreign opposition. The first item on his agenda is to crush dissidence and protest, whether peaceful or violent, and to muzzle press reporting of unrest. The goal is to establish order, or at least the semblance of order, so as to persuade foreign bankers and investors, as well as South Africans, that the government is in control of events. Only then will the government feel itself to be in a strong enough position to implement various measures for black political participation on terms acceptable to the white minority. Meanwhile it continues to ignore calls for negotiations with the black liberation movements.

There are of course grave doubts about whether this strategy will succeed – whether a calmer environment in the black communities will encourage moderate blacks to come forward and work in government-established bodies on which they have no real political power, or indeed whether the government's strategy and the forces at its disposal will be able to quell the country's unrest and create the breathing space it seeks. The current wave of protest, which began in September 1985, appeared to have peaked by mid-1986. But the troubles were by no means over.

Military and Diplomatic Moves in Angola

Early 1986 brought signs of change in both the US and the Communist world over policy towards Angola. In mid-February the Reagan Administration told Congress that it had decided to supply anti-aircraft and anti-tank missiles to Jonas Savimbi's UNITA guerrillas. This reversal of US policy of the previous ten years was a victory for senior White House staff, whose eagerness to apply the so-called 'Reagan Doctrine' to Angola prevailed over strong State Department reservations that the move could jeopardize regional peace talks and strengthen the Soviet position there.

Rumours that such a US decision was imminent had spurred Fidel Castro to make a strong pledge of support for the Dos Santos government in Angola, vowing to keep Cuban troops there until apartheid had been dismantled, if necessary. In February Angola, Cuba and the USSR signed a joint communiqué on solidarity, and by late March

South Africa reported that Luanda had recently received six MiG-23 interceptor aircraft and had deployed a large number of advanced Soviet radar and missile installations in its southern bases.

Between June and August heavy battles were reported along the central plateau, as government forces fought to pry UNITA guerrillas loose from their ten-year grip on central portions of the vital Benguela railway. As the rainy season approached in late October, however, the government offensive appeared to have stalled short of its objective, the town of Munhango. Apparently at the urging of Soviet and Cuban advisers, Luanda decided against launching an assault on UNITA's main stronghold in the country's south-east corner in 1986. But it does not seem to have abandoned the goal of dislodging Savimbi from his main base, and thus cutting his forces off from South African sources of supply. In an offensive in 1985 it had gained an important salient in the east (Cazomba), and during 1986 it reinforced its advance bases at Cazomba and Cuito Cuanavale in the south from which any future assaults on Savimbi's main stronghold would be launched.

South African military officials have expressed concern over Luanda's growing capability, both to carry out large-scale attacks against UNITA and to interfere with South African operations in support of the insurgents. The chief of the SAAF warned in a recent article that 'as the air umbrella [over Angola] has become more effective, it is . . . becoming increasingly difficult to neutralize it. And unless it is neutralized, no long-range operations in the host [sic] country are possible without heavy casualties'. Of particular concern to South Africa were the installation in 1986 of sophisticated ground control intercept radar around Cuito Cuanavale, and the deployment of Su-22s – advanced Soviet ground-attack aircraft – at the Angolan southern regional headquarters at Menongue. Among other new additions to Luanda's armoury are recent-model French and Soviet helicopters, including Mi-24 gunships to replace those lost in its 1985 offensive. Angolans now fly the helicopters and MiG-21s, but the more advanced aircraft, like MiG-23s, are said to be flown by Cuban and Soviet pilots.

If, as expected, Luanda decides to use its improved striking power in another major offensive against Jamba, UNITA's main base, the scale and intensity of fighting are likely to be far greater than in the recent past. The Angolan government now has a capability to provide air cover for assaults by its airborne and ground forces – a fatal omission in the 1985 offensive, when the SAAF attacked its armoured vehicles and helicopters virtually without opposition as they moved on Savimbi's forces. Any large-scale government offensive will almost certainly draw in the SADF, since UNITA's impressive guerrilla capabilities do not extend to defence against large-scale air and ground attack. That, in turn, would bring the risk of intervention by Cuban combat troops, who have in the past suffered heavy casualties in the fight against UNITA.

The war also threatened to spread to neighbouring countries in 1986. After UNITA initiated guerrilla attacks in Angola's northern enclave, Cabinda, in November, Luanda deployed troops along the border with Zaire, from whose territory it alleged that the attacks had been carried out. Zambia, too, came closer to involvement in the war when Savimbi accused it of allowing Angola to conduct assaults on UNITA bases from Zambian territory and threatened to carry the war into Zambia.

As 1987 began there were no signs of an early resolution of the conflict, which absorbs well over half of Angola's total budget and has virtually paralysed the economy. There have been unconfirmed reports that the failure to deal UNITA a crippling blow or even to curtail its operations may have weakened the government's commitment to a military solution. The intensity of its recent military preparations, however, suggests that it has not yet decided to give up the military option.

Mozambique: Daunting Problems

By early 1986 RENAMO guerrillas had recovered from their military reverses of the previous year and were carrying out sabotage in widely separated parts of the country. Only major military intervention by Zimbabwe, which has deployed some 12,000 troops in support of the government, kept open the vital corridor containing the railway and oil pipeline which run from Zimbabwe to Mozambique's port at Beira. In addition, 3,000 Tanzanian troops are helping to guard the pipeline, and in recent months Nigeria has promised to send 5,000 troops. But government forces – demoralized by poor pay, inadequate leadership and training, as well as food and equipment shortages – were reported to be deserting the army in large numbers; many appear to be FRELIMO by day and RENAMO by night.

By October, however, the government's counterinsurgency operations had cleared guerrillas from much of the south, particularly around Maputo, although 800 miles to the north RENAMO forces, operating from bases in neighbouring Malawi, had stepped up their operations during the year. In September President Machel and Presidents Kaunda of Zambia and Mugabe of Zimbabwe, made a joint démarche to Malawi, threatening to close its borders if it continued to support RENAMO. Malawi responded immediately by ordering all RENAMO forces to leave, which led to a major offensive, as some 10,000 heavily-armed guerrillas moved into Mozambique's upper Zambezi valley, capturing, and for the first time holding, major bridges and several towns. In November the government reported that Zambezia province was on the brink of collapse, with agriculture and mining brought to a standstill and an estimated four million people short of food and clothing.

South Africa continued to be ambivalent towards Mozambique. Senior officials on several occasions reiterated their support for the Machel government and the Nkomati Accord – the 1984 mutual security pact with Mozambique – but admitted that the SADF had violated it by maintaining clandestine contact with RENAMO.

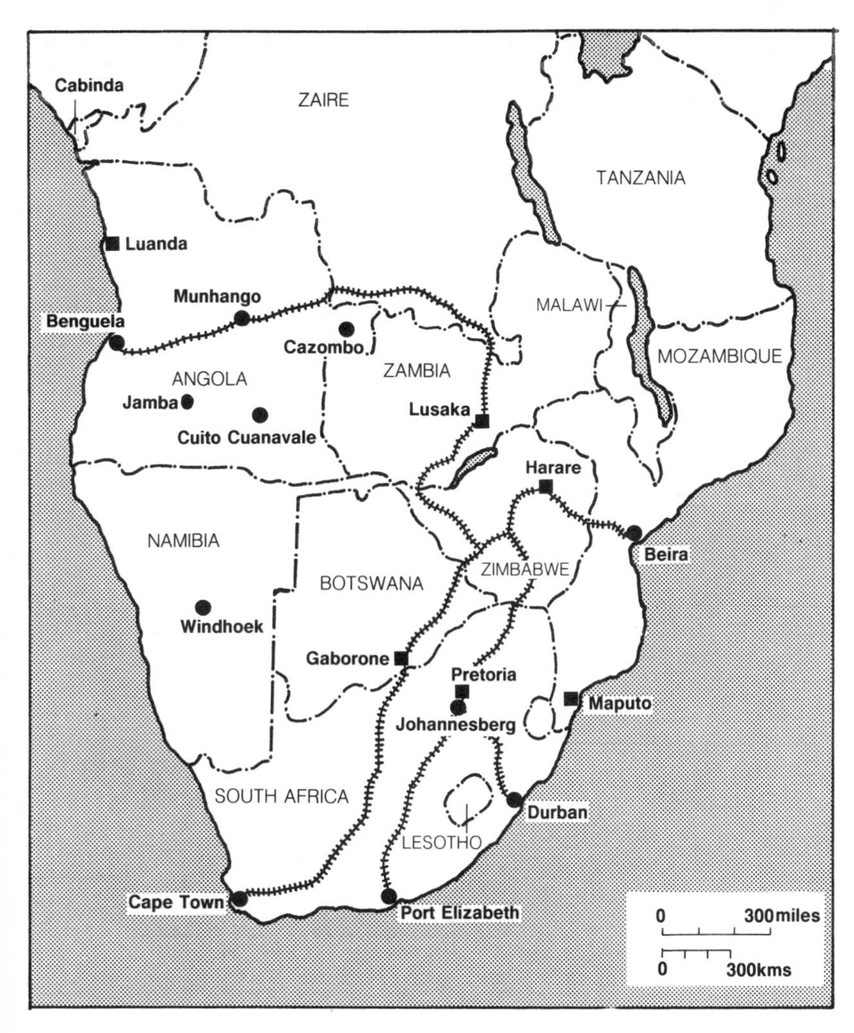

Western diplomats, including US Secretary of State Shultz, went further, accusing South Africa of continuing to aid RENAMO in 1986. In early October, less than a week after the US Senate overrode President Reagan's veto of the sanctions bill, South Africa announced that all the estimated 250,000 Mozambicans working in South Africa would be forced to leave when their current labour contracts expired. This move, which would cost Mozambique an estimated $50 million in foreign exchange receipts, indicated Pretoria's determination to retaliate against Western sanctions by punishing its neighbours – even those, like Mozambique, who had not called for sanctions.

On 19 October President Machel was killed in an air crash. The death of the charismatic leader who had led the FRELIMO forces in their struggle against the Portuguese was seen as a serious blow to morale in Mozambique and to the cohesion of the Front Line States' leadership. Within two weeks, however, the remaining leaders of the ruling FRELIMO party quietly selected Foreign Affairs Minister Joaquim Chissano to succeed Machel. Chissano, who is regarded as a pragmatic politician, reiterated Mozambique's commitment to non-alignment, regional co-operation and development, friendly ties with Communist countries and closer ties with the West, and a mixed economy. He also reaffirmed the government's commitment to building a socialist society and its determination to pursue the counterinsurgency operations against RENAMO which had consumed 42% of total budgetary outlays in 1986.

The immediate outlook for the new government was bleak. Starvation and guerrilla attacks had driven 300,000 Mozambicans to seek refuge in neighbouring states by the end of 1986, and another million faced starvation inside the country, as guerrilla actions continued to isolate large areas and paralyse normal trade between city and country in fuel, food and consumer goods. Yet Chissano had at least some grounds for optimism. Mugabe pledged that Zimbabwe would not allow a RENAMO victory and increased the number of Zimbabwe troops in Mozambique to an estimated 12,000. That they would take a more active role against RENAMO was indicated in November, when Zimbabwe asked for its troops to be allowed to cross Malawi to enter Mozambique's Zambezia province. It has now been established that before Machel's death in October Mozambique and Zimbabwe had been planning a coup against President Banda in an effort to rid Malawi of the RENAMO units which used the country as a sanctuary. With a 5,000-strong army of his own, Banda cannot police the border effectively.

Mozambique won expressions of diplomatic support from the US, Britain and Portugal during 1986, and in May the first group of 48 field-grade Mozambican officers completed a twelve-week training course run by Britain. Several foreign-aid donors rehabilitated the port of Maputo during the year, and there were further aid commitments to upgrade the Beira corridor and port. In March agreement was reached on joint ventures with private South African capital to run several hotels and other enterprises.

It remains to be seen whether the Chissano government will be able to turn the tide in the exhausting six-year war. Expulsion from its bases in Malawi will complicate RENAMO's acquisition of weapons, food and money from outside. Moreover, since it has neither the credentials of a national movement nor a political base among the populace, and has yet to offer a coherent political programme, its military successes have not been matched by political gains. Even so, and notwithstanding the intervention of Zimbabwe's reinforced and well-trained troop contingent, it is unlikely that in 1987 government forces will be able to do more than keep the southern region, the Beira corridor and major cities and ports clear of

RENAMO. Even with Tanzanian, Nigerian and Zimbabwean troops, the Beira corridor will be difficult to protect adequately. The rail link was cut three times early in 1987.

Problems in Zimbabwe and Zambia

The internal security situation in Zimbabwe improved during 1986, largely because of the continuing attempts at reconciliation between Joshua Nkomo and Prime Minister Robert Mugabe and the attempted unification of ZAPU and ZANU. Some ZAPU dissidents are still active in Matabeleland, but their activities have diminished markedly. There can be little doubt, however, that Zimbabwe is feeling extremely exposed on its southern border, because of Mugabe's hard-line stand on sanctions against South Africa.

At the Non-Aligned Movement's summit meeting in Harare in September, at which Mugabe was elected chairman of the Movement, a decision was taken to isolate South Africa and to render all possible military and financial assistance to the Front Line states. Indian prime minister Rajiv Gandhi undertook to provide military assistance to the Front Line states threatened by South African aggression and promised to provide Zimbabwe with aircraft, trainers and pilots (these have not yet been forthcoming, however, and Pakistan continues to train the Zimbabwean Air Force).

So far only $40 million in aid has been pledged by the Non-Aligned Movement. Meanwhile, Mugabe is believed to have lost the support of both his cabinet – and, more important, the ZANU Politburo – for his commitment to introduce sanctions against South Africa. A list of sanctions, promised in the wake of the Commonwealth mini-summit in mid-1986, was postponed until December and had not materialized by March 1987. Instead Zimbabwe is now focusing its hopes on Western aid for improving the capacity of the Beira rail link and port to reduce its dependence on South Africa. Currently 95% of its trade depends on the South African transport network.

In Zambia, the major difficulties remain the low state of the economy and a persistently high inflation rate (which touched 300% in 1986). Since the country is heavily dependent on the South African transport system, it is also, like Zimbabwe, extremely vulnerable to the possibility of counter-sanctions by Pretoria. (And President Kaunda, who led the calls for sanctions at the OAU meeting in 1986 and the Commonwealth mini-summit, has not yet announced Zambian sanctions against South Africa.) In the wake of the South African raid on alleged ANC bases in Lusaka on 19 May, there was increased sensitivity towards foreign tourists (several of whom were detained as suspected South African spies), and Zambia also continues to have problems on its porous Western border with Angola, which is frequently crossed by UNITA forces. In early December the government was rocked by widespread food riots after President Kaunda had been forced to drop food subsidies as a precondition for a badly-needed IMF loan. The riots, which were mainly in the Copperbelt area, were eventually put down by army units from outside the area.

South Africa Under Pressure

In the early months of 1986 the protest, disruption and violence in South Africa's black communities showed no signs of slackening, in spite of harsh actions by the security forces. By January some fifteen months of unrest had resulted in an official death toll of 955, of whom 628 had been killed by police action. While street gangs and self-appointed groups of vigilantes preyed on township residents, black organizations sponsored industrial strikes and peaceful (but illegal) consumer boycotts and rent strikes. The growing crisis led to greater polarization among whites, with the government's right-wing critics, in particular, showing increasing militancy and urging more extreme apartheid solutions to the country's troubles.

On 12 June President Botha re-imposed the State of Emergency he had lifted in March. Security forces rounded up hundreds of anti-apartheid activists under decrees that also curtailed the rights of detainees and of the press. One new decree authorized prosecution for making, printing or recording a subversive statement, defined as any remark 'weakening or undermining the confidence of the public'. Declaration of the emergency was timed to forestall a massive black protest strike on 16 June, the tenth anniversary of the Soweto riots, but in the Johannesburg and Cape Town areas the stay-away by black employees was nonetheless 90% effective.

Anti-apartheid leaders acknowledged in late August that the security crackdown was beginning to work. By then more than 8,500 people, including the upper echelons of black community and labour leadership, had been detained indefinitely without trial. Yet rent strikes spread to 42 black townships, and continuing school boycotts led to the closure of more than 75 black schools in Soweto and the Eastern Cape in October and November – although at the beginning of 1987 the children decided to end the boycott and return to school. In some black townships the vacuum left when local government representatives were forced to flee was filled by street committees which dispensed their own law: closing local drinking establishments and carrying out summary executions of alleged police informers and others.

Faced with continuing black unrest and an erosion of domestic and foreign confidence in the government's ability to surmount its troubles, the Botha leadership tightened security and press censorship even further in the final months of 1986. By the year-end more than 23,000 people had been jailed for varying periods under the June security decrees, and the SADF was patrolling the townships. The effect of these measures could not be judged because of the total ban on all but official reporting of black unrest.

The surge of black protest and government repression drew attention away from several limited but nonetheless important race reform measures adopted in 1986. Most notable was the termination of the hated pass laws and the restoration of South African citizenship to 1.75 million blacks; this was bad news, however, for residents of the nominally independent black homelands, who must in future apply for work permits as aliens. Other measures gave blacks freehold rights in black communities and removed the ban on black business in some white areas. There was, however, no move to rescind the Group Areas Act, which imposes residential segregation, although Botha announced in August that his government would follow a 'flexible' policy, turning a blind eye to the several thousand non-whites already living in 'white' neighbourhoods. The government also established new multi-racial Provincial Executive Committees to run regional affairs. Even moderate black leaders like Chief Buthelezi repudiated these bodies, however, because all the power rests with a white chairman appointed by the State President. Taken together, the government's piecemeal reforms attenuated some of the more glaring discrimination against blacks, but in no way weakened the political control of the white minority or of the National Party.

Worsening Relations with the West

For the first nine months of 1986 there was heated debate and political manoeuvring within Western Europe, the Commonwealth and the US over the issue of sanctions against South Africa. Discussions in the EEC were deadlocked as late as June, when Britain, West Germany and Portugal opposed a sanctions package proposed by the Netherlands. Among the Commonwealth states the Thatcher government was virtually alone in opposing stiffer sanctions than those adopted in 1985. In a compromise move, the Commonwealth had appointed an Eminent Persons Group (EPG) to explore the prospects for beginning a dialogue across racial lines in South Africa and to report in six months, when further measures would be considered. As a result of South African intransigence, the EPG cut short its mission in May and returned to London to issue a negative report which led to increased pressure for sanctions by Commonwealth leaders.

Meanwhile the Reagan Administration was fighting a rearguard action to fend off growing Congressional pressure for tougher US sanctions. In June it announced the start of a high-level policy review to explore ways of forging closer links with South African black leaders. However, despite harsher official rhetoric towards the Botha regime and hard lobbying in Congress, it was unable to prevent the Congress passing a new sanctions bill. The Administration offered last-minute concessions, including the appointment of a black ambassador to Pretoria and a large military aid package to the Front

Line states, but in October the Congress overrode the President's veto.

The US sanctions barred new corporate investment in South Africa and new loans to the government; banned imports of South African iron and steel, coal, uranium, textiles and some foods; terminated landing rights for South African Airways; and reimposed the milder sanctions adopted the year before, which had banned imports of Krugerrands and exports of nuclear power equipment and had restricted exports of computers. These were serious measures, which could affect almost a third of South Africa's commodity exports to the US.

In mid-September the EEC adopted similar, though weaker, sanctions. Because Germany vetoed a ban on imports of South African coal, the total package, including bans on purchases of gold coins and iron and steel, would affect no more than 5% of the Community's imports from South Africa. Prime Minister Thatcher remained strongly opposed to sanctions, but was bound by her promise to a delegation of Commonwealth heads of government in August that she would not prevent the EEC from imposing limited sanctions. In September Japan prohibited imports of South African iron and steel products, thus joining the country's other major trade partners in adopting at least token sanctions.

Economic Strains

Sanctions, serious enough at any time, were particularly unwelcome when South Africa was already suffering its fourth year of serious economic recession. A net capital outflow, totalling more than $2 billion in the last quarter of 1985 alone, continued in 1986 as investors demonstrated fading confidence in the troubled country. Foreign banks maintained their credit squeeze, and South Africa agreed to repay close to $2 billion of its foreign indebtedness in 1986. In July the head of the reserve bank said that, in the absence of 'comprehensive further political and constitutional reform', it was doubtful that the government's expansionary policies could prevent a siege economy, or stimulate economic growth any higher than 1.5–2% in 1986 – far short of the 4.5% needed to reverse the country's rising unemployment.

By the end of the year some 87 US firms had ended their South African operations – a few dissolving their South African interests; the majority engineering buy-outs by local managements or selling off their shares at a discount to South African companies. Most attributed their decision to the unfavourable economic outlook, but many also cited the 'hassle factor' – anti-apartheid pressures at home. More serious was the abrupt rise in white emigration: from a net gain of 5,896 in 1985, South Africa suffered a net loss of 6,200 in the first eleven months of 1986.

Not all the indicators were negative, however. Exchange reserves, augmented by a healthy export surplus and higher prices for gold and platinum, rose by more than $1 billion during the year, easing the strain of debt repayment. And in early 1987 the weakening dollar boosted the rand–dollar exchange rate by over 10%. The Reserve Bank remained cautiously optimistic that South Africa was 'poised' to start pulling out of the recession.

Will the Government's Strategy Work?

By late August, when government leaders decided against introducing legislation to create a multi-racial National Statutory Council, it was apparent that their twin-track strategy of repression and reform was running largely on the repression track, while reform had been temporarily derailed. Clearly the government is counting on two results from its strategy. First, that it will succeed in quelling black protest. Second, that a quieter political environment will be conducive to the acceptance of modest race reform from above by a majority of whites and blacks.

The success of the strategy will depend initially on the government's coercive capability. The South African Police (SAP) – the national police force – grew from around 35,000 to more than 40,000 in 1986 and is to reach 87,000 by 1990. Meanwhile training of auxiliary police has been accelerated: in September, after three weeks' training, a thousand black special constables were sent into black townships with full police powers, and another 6,000 municipal police were to be trained by the end of 1986. All members of the police now have virtual *carte blanche* to arrest and detain any individual and have been relieved of any legal liability for acts performed in good faith.

During the State of Emergency the army, too, has been called on to back up the SAP in patrolling the black townships, and recently it has also been given full responsibility for border security. The regular army, a small force of 18–20,000 professionals augmented by an annual influx of 50–60,000 white National Servicemen who flesh out its regular units, is backed by local Commandos and 140,000 active reservists. The government can thus muster, at relatively short notice, more than 200,000 well-trained, well-armed troops.

But not all of these forces would be immediately available to quell township unrest. Several thousand of those on active duty are serving in Namibia, and others patrol South Africa's borders with Botswana, Zimbabwe and Mozambique. Activity in the latter areas is likely to rise as the ANC attempts to carry out its recent threat to 'clear out' the border farmers, who are connected to the army communications network, and so provide a vital trip-wire against border infiltration. Moreover the army's large National Service component depends entirely on whites conscripted each year, and some white youths are balking at serving in the black townships – more than 7,500 of the January 1985 intake failed to report for military duty. Rising emigration, too, will to some extent reduce the limited white manpower pool on which conscription depends.

More germane is the question of effectiveness. The ANC leaders themselves 'do not foresee a situation now where [they] can mount military actions against the enemy' but will try instead to make the country ungovernable; they plan increased sabotage and attacks on those whites 'who are enforcing racist laws'. The government will continue to construct high-voltage fences and otherwise reinforce its border defences against guerrilla infiltration, but the sprawling black townships are likely to remain the focal point of black protest and unrest. Most of these ghettos are both isolated and distant from 'white' cities, so that it has been relatively easy for security forces to seal them off and, with a few exceptions, prevent the disturbances from spilling into white areas. Both the government and its opponents agree that the security crackdown has already dampened black dissidence inside the townships. All the same, the presence of troops in ghetto streets and schools had not brought rent boycotts or disturbances in the black schools to an end.

Two years of protest and repression, with widespread violence by both dissidents and police, appear to have further polarized the country's blacks (and whites). Opinion polls suggest that most blacks remain opposed to violence, and many would like to see an end to the troubles which have made life in the townships even more dangerous and uncertain than in the past. Yet the protest and repression have led to a rise in the number of black political activists and a wider acceptance of violence as a tool for bringing about change. Inevitably, new black protest leaders will emerge to take the place of those in jail or exile. In early 1987 the most likely prospect appeared to be that unrest will be further dampened by the exercise of overwhelming force, but that it will continue to simmer just beneath the surface, ready to burst out again at any time.

Meanwhile President Botha has called for a white general election on 6 May 1987. What he seeks is a substantial vote of confidence from whites in his policy of repression and modest race reform from above. That, together with a drop in overt black unrest, would encourage him to try again to tempt moderate black leaders into serving on government-created bodies that have been carefully designed to avoid giving blacks any real political power. So far even moderate blacks have rejected these institutions. It is highly doubtful that any mix of race reform measures can be found that would simultaneously satisfy most blacks' demands for real power-sharing and most whites' insistence that the reins of power be kept in white hands.

Latin America

CENTRAL AMERICA: CONTINUED DEADLOCK

The outlook in Central America remains unpromising. The past twelve months have seen little progress towards stabilizing the continuing crisis, no success in further attempts to secure peace through talks either within El Salvador or between the five Central American countries, and a number of developments – in particular the authorization of US military assistance to the Nicaraguan 'Contra' rebels – which threatened to cause a further escalation of the conflicts. Despite this grim news, there were some advances for Democracy. In Costa Rica, where there is a long tradition of political democracy, elections in February brought another peaceful change of administration. The inauguration on 14 January 1986 of President José Azcona in Honduras was the first peaceful handover of power from one government to another in that country's history. And Christian Democrat Vinicio Cerezo's inauguration on 14 January 1986 brought a moderate civilian to the Guatemalan presidency after decades of almost unbroken military rule.

There are still serious problems in all these countries, however. Particularly in Guatemala, despite Cerezo's achievements, much must be done to demilitarize the nation and reorient its development before the country will be firmly on the path of social progress and stable civilian democracy. So long as the wars in El Salvador and Nicaragua continue and the region remains in a state of tension and conflict, moreover, it will be all the harder to resolve these and all the other basic problems which contribute to the area's turmoil.

El Salvador

After another year of war and economic crisis, the country is no nearer to peace. During 1986 the guerrilla forces of the Farabundo Martí National Liberation Front (FMLN), whose numbers are now variously estimated at between 5,000 and 9,000, continued to operate in small units, concentrating on economic sabotage and disruption in a war of attrition intended to destabilize the government and force it into political negotiations. They can still mount occasional large-scale attacks, and they can certainly operate throughout wide tracts of the country, but the areas in which they exert some control appear to be shrinking, and they continue to have problems of supply.

The armed forces, which together with the security forces number over 50,000, are now trying to consolidate the initiative gained as a result of their improved equipment, mobility and performance, and to translate it into durable control in the countryside. A broad new plan, 'United to Rebuild', was presented (by army Chief-of-Staff Gen. Adolfo Blandon, rather than by the government) in August. The plan – said by Gen. Blandon to respond to the fact that the guerrillas' strategy was only 10% military, but 90% political, economic, social and ideological – consists of phases of mopping-up operations, consolidation and reconstruction in co-operation not only with the government but also with the private sector. Whether this plan is more successful than its predecessors will largely depend on whether the government forces can convince the population that they are now acting in the interests of democratic rule and social improvement.

No massacres by government forces have been reported for over two years and the number of civilian casualties from aerial bombing appears to have fallen. But abuses still take place. Political killings, 'disappearances' and incidents of physical torture are much fewer than in the past, but they have not ceased, and serious mistreatment of prisoners and suspects appears to be frequent. However, the guerrillas may not be able to take full advantage of this, for their own behaviour tends to reduce their support as well. Their increasing reliance on mines has led to significant civilian casualties, and they have carried out many political killings, attacks on traffic and kidnappings.

Peace negotiations came to nothing in 1986; the gulf between the two sides, which had been made so clear at the second round of talks in November 1984, seems to be unbridgeable. A meeting on 26 April in Lima between a senior Christian Democrat and two moderate leaders of the Revolutionary Democratic Front (FDR), the political wing of the revolutionary alliance, turned out to be fruitless; nonetheless, on 1 June President Duarte, under strong union pressure for peace talks, announced his willingness to start the process again. In reply, the FDR/FMLN presented a new, modified peace proposal on 14 July. It called for a cease-fire, the formation of a broad coalition and the subsequent holding of new elections, but did not include a merger of the two sides' armed forces, which would remain intact until the 'process of democratization' (including a purge of the government forces), was completed. To counter objections that the formation of a provisional government would be unconstitutional, it also proposed not a coalition government but a coalition cabinet. Duarte, however, rejected the plan as not being substantially different from former open demands for a power-sharing solution.

Talks in Mexico on 20–23 August produced a tentative agreement to begin a third round of talks on 19 September in the small Salvadorean town of Sesori. During preparatory talks in Panama on 12–14 September even that agreement collapsed, because the government refused to accept rebel demands for a cease-fire and the withdrawal of all government forces from the Sesori area. Although both sides insisted that this failure did not mark the end of the negotiating process, the prospects for any serious dialogue seemed bleaker than ever.

The military initiative still appears to lie with the government forces, and that trend seems likely to continue. Yet the guerrillas, weakened though they are, will be able to keep up a significant level of activity for some time. Moreover, the centrist government under the Christian Democrat President Duarte – the establishment of which has been central in securing solid US backing, increased international legitimacy and greater domestic support for the current political process – seems destined to be progressively weakened the longer the war goes on.

Because it has been unable to create either peace or social and economic improvements, the government has continued to lose popular support and credibility in the country – a phenomenon best seen in increasing labour protests, even by those union groupings which had previously been supportive. It has also come into greater and more open confrontation with the political Right and the business sector, which has strongly attacked the government's management of the economy. A package of adjustment measures (including price rises for many basic commodities) brought labour protest and mobilization at the beginning of 1986, while in the first days of 1987 attempts to increase taxation on business and wealth evoked fury from the the private sector, a boycott of the National Assembly by right-wing deputies and some calls for a military coup. Although Duarte will probably survive until the end of his term in office, it seems increasingly unlikely that the Christian Democrats will be re-elected. If this weakening of the centre leads to renewed political polarization, there could be a significant setback for the current political process and a considerable resurgence of the conflict in El Salvador.

Nicaragua: Constitution, Contadora, Contras

In the midst of war and economic crisis, the process of political institutionalization continued in Nicaragua, centring on the preparation and approval of a new national Constitution. In February the Constitutional Commission presented a draft text to be submitted to popular discussion in 'open fora' in May and June, and then debated in the National Assembly from September. Although the draft contained most of the principles of representative democracy, a number of national parties and sectors had indicated their scepticism by opting out of the process. The conservative Nicaraguan Democratic Co-ordinating Council had boycotted the November 1984 election; the Independent Liberal Party had withdrawn from the preparatory Commission after the tightening of the State of Emergency in October 1985; and in April 1986 the Democratic Conservative Party threatened not to take part in any of the remaining stages of the discussions.

The opposition's objections were partly related to specific elements in the draft text: the role of the armed forces; the provisions dealing with the mixed economy and private property, considered to give the state an excessive role; the overwhelming power given to the executive branch; an insufficiently clear separation of the ruling party from the state; and the alarmingly vague qualifications that accompanied many fundamental provisions for political pluralism, electoral transfer of power, civil liberties and a mixed economy. More broadly, it was felt that the restrictions on political activity – justified by the Sandinistas as necessary in view of the war – made democratic debate impossible, while some argued that discussion of a constitution was irrelevant so long as the other parties lacked the confidence that the Sandinista Front would abide by it.

Attempts at a Peaceful Solution

A similar pattern of distrust was discernible in relations among the Central American countries. A summit meeting of the five presidents was held in Esquipulas, Guatemala, on 24–25 May. They agreed to create a directly elected Central American parliament as a means to strengthen 'dialogue, joint development, democracy and pluralism as the key elements for peace'. Although the depth of political suspicion was evident, hopes were raised that this approach might create an opening for a more favourable climate for both the promotion of democracy in all the countries and the peaceful resolution of security concerns.

In early April the Contadora Foreign Ministers had urged the five countries to complete their negotiations in order to sign an Act for Peace and Co-operation on 6 June. During May, however, differences over international military manoeuvres and arms limits remained unresolved. Nicaragua, anxious to remove the threatening US military presence in Honduras, insisted that all exercises in progress be halted within thirty days of the signature of an agreement, and that all future exercises be entirely prohibited; the other countries proposed only to limit them. With regard to arms limitation, the other Central American countries insisted that limits should be agreed before signature and should be based on such objective criteria as a country's size, population and gross domestic product. Nicaragua wanted negotiations for arms limits after signature, with emphasis given in defining them to a country's own perceptions of its security needs.

Although there were attempts at compromise, they were blocked by the lack of political trust among the Central American governments themselves. At least as important, however, was the role of the US. The problem was not simply that of 'simultaneity', for on this issue there appeared to be signs of flexibility on both sides. With the Contras greatly weakened and the US vote on Contra military aid approaching, the Sandinistas ceased to emphasize the need for a prior renunciation of US hostilities. On the other hand, the US special envoy, Philip Habib, seemed to indicate that the US would be prepared to end support for the Contras at the same time as an agreement acceptable to all the Central American countries was signed.

So long as 'national reconciliation' within Nicaragua was understood to mean political negotiations with both unarmed and armed opposition, which the Sandinistas have rejected, any settlement could be predicted not to resolve the confrontation. In addition, the Reagan Administration suspected that the Sandinistas might not comply with even the security provisions of a settlement and was thus reluctant to withdraw its pressures on Nicaragua. The Sandinistas, for their part, suspected that the US would not respect any regional settlement which left them in power.

Sandinista suspicions seemed to be confirmed by the outcry from US conservatives after Habib's apparent indication of willingness to settle with Managua, by the release of a Pentagon document emphasizing the alleged danger for both regional and US security if the Contadora Act were to be signed, and by open White House statements that the Administration would continue to support the Contras even if a settlement were signed. This promise of continued military pressure, supposedly to ensure that all elements of a settlement were actually implemented, was predictably matched by Sandinista insistence that negotiations for a reduction in military forces should take place only after a settlement was signed, supposedly to ensure that the US would in fact respect the agreement. In this context, the prospects of the Central American countries overcoming their differences were practically non-existent. The 6 June deadline was not met.

At their meeting in Panama on 6–7 June the Foreign Ministers of the Contadora Group and Support Group issued a joint declaration which was long on pious appeals but sort on specifics. It called on all countries in the area to respect each other's sovereignty, not to join military or political alliances that would involve the region in the East–West conflict, not to commit aggression, and not to support irregular forces or subversive groups to overthrow a government in the area. The new, supposedly final, version of the Peace Act which was presented was considered much too favourable to Nicaragua by its neighbours and quickly rejected.

The failure of the latest Contadora initiative and the apparent lack of any increase in consensus in Nicaragua contributed to the US authorization of military aid to the Contras. This in turn contributed to the further sharpening of tension which was detectable in developments in Nicaragua and Central America in the following months. Since March the Reagan Administration had been pressing on with its campaign to gain Congressional approval of a programme of assistance to the Contras involving $US 70 million in military aid and $US 30 million in non-military assistance. Recent deliveries of Soviet military equipment to Nicaragua, increasing domestic repression by the Sandinistas, and in May an agreement among Contra leaders appearing to give greater control to moderate civilians added weight to the Administration requests. The package was eventually approved in the House on 25 June by the narrow margin of 221 to 209. Of the $US 100-million package, $US 60 million was approved immediately and the rest reserved for a further vote in March 1987.

The Sandinistas' response to the June vote, which was seen as a declaration of war, was quick and strong. The following day President Daniel Ortega announced that the State of Emergency would henceforth be strictly applied, the only opposition newspaper, *La Prensa*, was closed indefinitely, and opposition activity was further restricted. Actions against Church leaders who had taken an active anti-government stand began on 28 June with the refusal to allow the spokesman of the Nicaraguan Episcopal Conference to re-enter the country. On 4 July the government expelled Bishop Antonio Vega for declarations said to support military assistance to the Contras. It also was announced that there would be increased conscription and mobilization into the reserves.

These actions predictably sharpened political tensions in Nicaragua. In early July opposition parties, most of which had not participated in the 1984 elections, called for the formation of a National Peace Commission by all Nicaraguan civilian groups, the revocation of the new measures, and the subsequent holding of new elections. On 1 September five of the opposition parties in the National Assembly requested that the debate on the proposed new Constitution should be suspended for two months and called for a 'national dialogue', to include those parties which did not have legislative representation. The government responded by announcing that opposition parties would be permitted to hold political meetings and to publicize their views for the period of the Constitutional debate. The five parties nevertheless threatened not to participate in the debate unless some political agreement was first reached, and half the opposition representatives stayed away from the ceremony on 9 September at which the Constitutional draft was presented to the National Assembly.

Some easing of tensions seemed to take place in the autumn. As a result of mediation by the Papal Nuncio, delegations from the Episcopal Conference and the government met on 27 September to try to normalize relations, and a mixed government/episcopal commission held two meetings. At the same time the legal opposition parties returned to the final debates on the Constitution, which was finally approved on 19 November. However, the dialogue with the Church had not brought any practical results by the end of the year, and significant fractions of both the Democratic Conservative and the Independent Liberal parties refused even to sign the Constitution, which came into force on 9 January 1987.

At the same time, negotiations among the Central American countries became more difficult. Following their rejection of the fourth Contadora draft, El Salvador, Honduras and Costa Rica decided to present an alternative peace proposal. The Sandinistas, encouraged by the World Court's final ruling against the US for its actions towards Nicaragua, brought similar cases on 30 July against Honduras and Costa Rica for their alleged support of Contra forces.

Costa Rica, a traditional supporter of international law, stated that it would fight the case. Honduras, following the US lead, announced that it did not recognize the World Court's jurisdiction. Both governments were angered by the move and refused to participate in further Contadora discussions unless Nicaragua withdrew its case before the World Court. The same effectively applied to further discussions being held on the question of a Central American parliament, and these ceased in October. A new Contadora initiative began to be prepared in December, but the January 1987 mission to Central America by the eight Foreign Ministers of the Contadora and Support Groups and the Secretaries General of both the United Nations and the Organization of American States was stymied on familiar grounds. Costa Rica and Honduras, supported by El Salvador, continued to insist that no settlement would be reached unless it contained provisions for democracy in Nicaragua; and Nicaragua continued to insist that support for the Contras had to stop.

Weakening Support for the Contras?

Even as the situation in and around Nicaragua thus appeared to become more volatile and intractable, other developments had begun to make a change seem possible in one of the key elements in the present impasse: namely the continuation of the US-backed Contra war.

On 5 October 1986, an aircraft carrying military supplies to be dropped to Contra forces operating inside Nicaragua was shot down by Sandinista forces. Statements by the one American survivor about this 'private' flight confirmed that it had originated from the Ilopango military air base in San Salvador and that the aircraft was owned by a company that had long been used by the CIA for operations. The revelations confirmed the very thin line between private aid to the Contras and military support involving government figures and agencies at a time when official assistance was legally limited to the 'humanitarian' aid authorized by Congress in June 1985.

On 4 November the US elections gave the Democratic Party control in the Senate as well as the House of Representatives, thus making opposition to future Contra aid more likely, although not certain. Then came the extremely damaging scandal over the apparent illegal diversion to the Contras of funds from secret arms sales to Iran. It now seemed very likely that – even though Congress agreed in March 1987 to release the final $US 40-million tranche of the aid package approved the previous summer – the President would have great difficulty in securing its approval of the further $US 150 million requested for aid to the Contras in his State of the Union message in early January 1987.

This situation came on top of, and clearly reinforced, the growing doubt and discomfort about the Contras which Nicaragua's neighbours had felt even as tensions with the Sandinistas had been rising. The new Costa Rican government of President Oscar Arias made no secret of its political dislike and distrust of the Sandinistas, but insisted that neutral Costa Rica would not be involved in armed activity against the Nicaraguan government. Despite apparent US requests and pressures, it refused to support the establishment of the much-needed southern front, and took various steps to counter military use of Costa Rican territory, including the eventual closure of an airstrip allegedly used to supply Contra forces.

The Honduran government had continued to insist that military aid would not be provided to the Contras in Honduras, and the Contra presence had been becoming more and more openly and broadly unwelcome through the year. Not only was there concern about the country's role and image and about more specific border problems, but the Contra presence was increasingly disrupting civil life and agricultural production, and even displacing the local inhabitants. Serious border incidents, coinciding with the scandal in the US, seemed to bring Honduran concerns to a head.

Tensions escalated in November as Sandinista forces deployed in strength to prevent Contra infiltration and followed through with a number of relatively large-scale incursions into Honduras in pursuit of Contra units. In early December the Honduran forces, which until then had stayed clear of the fighting, came into direct contact with Sandinista troops, and responded by striking targets inside Nicaragua with A-37 ground-attack aircraft. Thereupon the US – to demonstrate both its commitment to Honduras and the supposed dangers of Sandinista rule at a difficult moment for US policy – provided well advertised logistic support to the Honduran forces, as it had done in similar circumstances in March. (In November, it had already publicized a decision to accede to a long-standing request to sell Honduras either American F-5E or Israeli *Kfir* jet fighters to replace the squadron of aged *Super Mystères* which were its principal means of countering Nicaragua's much larger ground forces.)

The incident, however, was a painful demonstration to Honduras of just how isolated it now was. It had received practically no external support, other than that from the US. Internally there was a general lack of confidence in the Contras' prospects, especially while they were staying in Honduras attracting Nicaraguan attacks, and doubts about the prospects for the US Administration's policy were growing. In mid-December, the government reportedly told the US Ambassador in Tegucigalpa that the Contra forces should leave Honduras in April and move into Nicaragua, where they should be fighting if they were to have any prospect of success, and where Honduras would prefer them to be if their vital support were to be cut off or they were otherwise to approach collapse.

Even with US support, the Contras are very unlikely to be able to take and hold any territory in Nicaragua where they could establish effective bases. Indeed, with no major offensive mounted since October 1985, there were perhaps as few as 2,000 Contra combatants still in the country at the beginning of 1987, concentrated in one central region and attacking only lightly defended or purely civilian targets. The Sandinista regime's control over the northern areas has improved greatly, and, with improved use of its fleet of Soviet helicopters and the deployment of highly mobile irregular 'hunter' units as well as irregular warfare battalions, its overall military capabilities have increased. The Contra forces are not finished, and will probably be able to continue their campaign for some time, if some external support continues. Indeed, in view of their urgent need to demonstrate some military credibility, both to the Nicaraguan population and to the US Congress, it seems likely that they will soon try to carry out some more impressive actions. But, even with US aid, their prospects of making any durable progress still look limited, and it seems increasingly likely that the Sandinista regime, despite its many weaknesses and vulnerabilities, will be able to hold out and consolidate itself – unless the US opts for much more direct involvement in the war.

Any Congressional decision to withdraw support from the Contras will certainly depend upon whether there are also significant signs of change at the local level which promise genuine progress towards regional peace and democracy. Movement in the stalled Contadora process would also seem to depend upon the withdrawal of Nicaragua's World Court case – more likely, and politically easier, for Nicaragua if Honduras and Costa Rica were seen to act on their stated desire to end Contra use of their territory – as well as on all countries' willingness to accept a mutually satisfactory compromise in local terms. Progress towards political understanding within Nicaragua, which the Constitutional process had not ruled out, will clearly also depend upon whether the Sandinistas can demonstrate some real willingness to move towards full pluralist democracy and non-alignment.

The prospects for Central America in 1987 are thus far from clear. US support for the Contras may be continued, which at best would practically guarantee a perpetuation of the current situation. However, if that support is withdrawn by Congress, it is then possible (but highly unlikely) that more direct action against Nicaragua could still be taken by the Administration, or precipitated by events. The most probable course, however, is that a weakening of support for the Contras, particularly in Washington, might create another opportunity for all the parties to begin to break out of the continuing Central American deadlock.

SOUTH AMERICA: DEBT UNDERMINING DEMOCRACY

The twin themes of debt and democracy, which have dominated the South American political scene for the past few years, still held sway during 1986. Although there was no major political breakdown in the region, and the debt crisis continued to be 'managed', the pattern of events left little room for optimism. On the one hand, the overall economic situation had seriously worsened by early 1987, and the traditional techniques of debt management had come under increasing strain. On the other, the political difficulties of consolidating democratic regimes in such a hostile environment had intensified.

The Economic Constraints

Although the overall economic situation could not be considered anything but gloomy, there were at least a few encouraging signs during the year. In the first place, figures from the UN Economic Commission for Latin America and the Caribbean showed that Latin America's GDP rose by 3.4% in 1986 (against 2.7% in 1985), with those of Peru and Brazil growing by 8% and those of Argentina, Chile, Colombia and Uruguay by between 5% and 6%. Secondly, three major countries adopted and implemented new domestic economic strategies which achieved some success in reducing inflation (Argentina with the Austral Plan, Brazil with the Cruzado Plan, and Peru with its Peruvian Plan). These plans were basically short-term emergency responses to the twin problems of inflation and stagnation and involved the freezing of exchange rates, strict price and wage controls, a lowering of domestic interest rates in line with inflation and the creation of new units of currency. In Argentina inflation fell from 385% in November 1985 to 86% in November 1986, and Peru's annual inflation rate dropped from 192% to 60% over the same period. In Brazil monthly inflation fell from 13% in February 1986 to an average of 1.5% over the the period from March to October. In Latin America as a whole, inflation fell in 12 out of 22 countries, with the average population-weighted rate of increase dropping from 275% in 1985 to less than 70% at the end of 1986.

Thirdly, the continued fall in international interest rates benefited all the region's major debtors. By the second half of 1986 base interest rates in the international market stood at their lowest level since 1977, saving the region some $US 5 billion in interest payments. Finally, although the fall in oil prices had a disastrous impact on some countries (above all Mexico), it provided substantial benefit to the region's oil importers. The cost of Brazil's oil imports, for example, fell from $US 9.8 billion in 1985 to $US 5.2 billion in 1986.

Despite these positive signs, 1986 brought little general relief from the severe economic constraints that have dogged the region since the onset of the debt crisis in 1982. The external environment remained unfavourable, with the fall in international oil prices and the continued low level of most commodity prices reducing export earnings for the region as a whole. Latin American exports declined in value by 15% in 1986 (as against 6% in 1985), while those of the oil-producing states fell by 34%. Coupled with a further moderate recovery in imports, the region's overall trade surplus dropped from a record $US 39.4 in 1984 to $US 18.5 billion in 1986. There was thus a sharp decline in its ability to generate the trade surpluses necessary to finance interest payments, and the overall trade surplus in 1986 was equivalent to 47% of total interest payments, as against 80% in 1985 and 89% in 1984.

Moreover, what gains there were in 1986 were not generalized, and there was a major distinction between the performance of the oil-producing and non-oil-producing states. Thus Brazil's and Peru's high growth rates need to be set against figures of only 1.5% for Venezuela and Ecuador, whilst Mexico's economic situation was still more serious, registering falls in GDP and industrial production of 4% and 6% respectively, a fiscal deficit equal to 15% of GDP, and annual inflation of around 100% (up from 60% in 1985). Finally, by the end of the year some of the apparent gains were wearing thin. In Argentina, for example, inflation rose steadily through the second half of 1986, reaching a monthly figure of 7.6% in January 1987 – against a target of 3% agreed with the IMF. In Brazil monthly inflation was 16% in January 1987 (higher than before the introduction of the Cruzado Plan), interest rates had soared to over 600%, and the country's trade surplus had shrunk from a monthly average of $US 1 billion for most of 1986 to an average of just $US 165 million in the last quarter.

Consolidating Democracy

Against this dismal economic background the problems of consolidating the civilian governments that had emerged from the wave of democratic transitions during the early 1980s were multiplied. All these governments faced a daunting double challenge: to develop political and constitutional structures that could accommodate the new social forces suppressed under the preceding military regimes, and to pursue economic policies that could provide both economic stabilization and socially acceptable levels of growth and development. Measured against the scale of the challenge, a degree of success should be noted. The gloomy predictions of political instability and mass social unrest expressed since the early days of the debt crisis have still not been fulfilled. Political structures in general proved resilient enough to cope with continued painful economic adjustments. And the new generation of moderate leaders – José Sarney in Brazil, Raul Alfonsín in Argentina and Alán García in Peru – did not face any decisive challenge in 1986.

García's personal popularity – based on his populist appeal, his youth and dynamism, and his radical image – remained strong. This popularity and his success in extending the organizational strength of the American Popular Revolutionary Alliance (APRA) party beyond the northern coastal area was confirmed by the results of the November municipal elections. In Argentina, too, democracy appeared to be well established. Alfonsín's own popularity remained high, the parties of the far right and far left had been effectively confined to the margins of politics, and the power of the military and the level of military spending had been curbed. Even in Mexico, reports of the imminent collapse of the dominance of the political system by the Institutional Revolutionary Party (PRI) seemed to be premature.

Nevertheless, despite the achievement of formal democracy and the absence of widespread social unrest, serious problems faced the region's political leaders. The stability of Alfonsín's political position in Argentina needs to be set against his failure to deal with the two major issues facing the country: incorporating the trade unions into a national economic consensus and overhauling and reforming the archaic public sector. In Peru the guerrilla challenge from the ultra-left-wing *Sendero Luminoso* (Shining Path) continued, as the group extended its activities into the slums surrounding Lima. Moreover, the bloody suppression of mutinies in three Lima jails in June highlighted the divergence between García and the military over how this threat should be tackled.

The rapid deterioration of Brazil's economic and political fortunes in 1986 graphically illustrated the volatility of the situation. For much of the year there had been grounds for cautious optimism. The country's economy was growing at around 8% p.a., its trade surplus for the first six months totalled $US 6.1 billion, and the Cruzado Plan introduced in February had dramatically cut the rate of inflation. And politically, despite the fact that President Sarney was an establishment figure from the military period who had never expected to become president, he attained extremely high levels of personal popularity. This was based, above all, on the economic successes of the Cruzado Plan, but also on his nationalist stance on the debt question and his determination to pay much greater attention to what he termed Brazil's 'social debt' – the legacy of mass social deprivation and inequality inherited from the military period and before. This popularity bore fruit when the Brazilian Democratic Movement (PMDB), the major party in the government coalition, won a landslide victory in the 15 November elections, taking 22 out of 23 state governorships and winning an absolute majority in both Houses of Congress.

One week after the elections, however, the government was forced to adopt an extremely tough package of economic measures to reduce the overheating in the economy. These adjustments – above all the removal of the freeze on prices – had been postponed until after the elections for political reasons, and this led to widespread feelings of resentment amongst the government's supporters and to rioting in Brasilia, Manaus and Rio de Janeiro. Moreover, the adjustments failed to counter the resurgence of inflation. Thus by early 1987 Sarney was no longer dominating the political scene. Apparently unable to implement a coherent economic policy, he seemed increasingly isolated from his political base, and the PMDB showed itself to be determined to play a more assertive and autonomous role in the deliberations of the Constituent Assembly which in early March began debating the new constitution that will determine the future shape of the Brazilian political system (including new laws on foreign investment and the length of Sarney's own term of office).

In the most important country still under military rule, Chile, the political crisis deepened in 1986, with the attempted assassination of Gen. Pinochet in September and the hard-line response of the military illustrating the dangers of polarization. Although the opposition had hoped that 1986 would prove to be a decisive year in the struggle for power, the uneasy stalemate continued.

On the one hand, the opposition to Pinochet continued to grow. The alliance of eleven political parties, formed in mid-1985 around the 'National Agreement' but dogged by political infighting, was reinforced by the Civic Assembly established at the end of April. The Assembly brought together a wide range of professional associations, unions and other organizations, and its capacity to mobilize broad support was manifested in the mass demonstrations of early July. In addition, there were unusual signs of disunity in the military, which was illustrated when the heads of the Navy, Admiral Merino, and the Air Force, Gen. Matthei, criticized Pinochet's July statement that he intended to remain as head of state until 1997 and his refusal in September to hold a plebiscite on his anti-terrorist policies. A further sign of disunity was the open criticism of the actions of the Army's Fifth Division, based in Punta Arenas, over the murder of a prominent left-wing journalist. Finally, there was growing, though not yet directly effective, US pressure on Pinochet to speed up the move towards democracy.

On the other hand, Pinochet's determination to remain in office and the significant political power that he still has at his disposal were clear to see. The reintroduction of the state of siege in September demonstrated his determination to continue using repressive tactics. The replacement of Gen. Canessa on the four-man junta by Gen. Humberto Gordon, director of the secret police, gave Pinochet closer control over the highly effective security apparatus centralized in that body and strengthened his position within the military. Finally, and most importantly, the democratic opposition remained disunited and uncertain. By October the Christian Democrats had effectively withdrawn from the Civic Assembly, and there was growing opposition within the moderate Socialist Party towards the Communist-led Popular Democratic Movement. By the end of the year, therefore, the opposition's ability to mobilize the middle ground of Chilean politics against the military remained in doubt.

External Relations

For those countries not closely involved in the crisis in Central America, the question of debt management dominated Latin American international relations in 1986. Two trends emerged: an increasingly tough stance was being adopted by the debtor nations, and the framework of debt management that had successfully contained the crisis since 1982 was becoming further eroded. The second trend had begun in 1985, with Peru's decision to limit interest payments to 10% of export earnings and Brazil's refusal to submit to an IMF programme. In 1986 it was Mexico that was at the forefront. In July, after protracted negotiations, an agreement was reached in principle (it was finalized in September), under which Mexico would receive a total of $US 12 billion of new money in the remaining eighteen months of the De La Madrid presidency, half of it coming from commercial banks, the rest from governments and the IMF. In addition, Mexico won three important concessions: the interest charged on the $US 44 billion of debt rescheduled in 1986 was reduced; the IMF took a more flexible approach to Mexico's fiscal deficit; and, most importantly, the country's creditors pledged that further funding would automatically be made available if growth in 1987 fell below 3%–4% or if recovery was impaired by some external shock, such as a further fall in the oil price.

In early 1987, while the banking community was still having difficulty in persuading the smaller banks to subscribe to the Mexican loan, Brazil delivered two further challenges. In January it successfully threatened to suspend negotiations with the Paris Club group of commercial banks over rescheduling $US 4 billion of official debt unless the creditors dropped their demand for formal IMF monitoring of the Brazilian economy. Then, on 20 February, it announced an indefinite suspension of interest payments on its $US 68-billion long-term private debt – a move which undoubtedly marks a significant change in the pattern of debt negotiations.

The debt crisis once more overshadowed relations between Washington and the major Latin American states, and the deep divergence of priorities was obvious. Washington continued to focus its regional policy primarily on the fight against Communism in Central America, arguing that the debt crisis should be resolved between the debtors and their private creditors. But even moderate Latin American leaders became increasingly frustrated by its refusal to treat the debt as a government-to-government issue and by the evident failure of short-term rescue packages to provide a viable long-term solution. This divergence was well illustrated by President Sarney's visit to Washington in September, which failed to resolve the protracted and bitter trade dispute between the two countries over Brazil's restrictive computer law which bans foreign companies from producing mini-computers in Brazil.

One important shift in US policy concerned Chile and the stepping up of pressure on Pinochet to ensure an early return to civilian rule. Although Washington had openly supported such a return since 1983, it had hitherto argued that its leverage over Chile was limited by the absence of bilateral economic and military aid, and that it was unwilling to use its power in international financial bodies to influence the internal politics of debtor nations. However, pressure on Chile was increased in 1986 in several ways: President Reagan made several speeches criticizing Chile, Nicaragua and Cuba as the region's only remaining dictatorships; in March 1986 the US supported a United Nations report on human rights that was highly critical of Chile; there were increased contacts between Washington and the civilian opposition; and, above all, the US threatened in July to vote against proposed loans to Chile by the World Bank and the Inter-American Development Bank. In the end, the US Administration abstained in the votes on the loans, but early 1987 saw renewed signs of its determination to tighten the pressure on Chile.

Within the region, two trends were noteworthy. In the first place, although there was no formal debtors' cartel, the level of formal and informal contact and co-operation between the major debtors reached a high level. Secondly, 1986 saw a solidification of the rapprochement between Brazil and Argentina that had begun in 1979. Interest centred on the two trade agreements signed by Presidents Sarney and Alfonsín in July and September, but there was also significant progress in the fields of computer technology, joint aircraft production, energy and the nuclear sector. Although serious obstacles to greater economic integration remain, the continuation of improved relations underlines a major change that has taken place in the pattern of Latin American international politics.

Beyond the hemisphere, attempts by Latin American states to broaden their political and economic relations and to recover the ground lost in the early 1980s continued to be hampered by the caution displayed by Japan and Western Europe and by the economic weakness of much of the Third World. There were, however, two significant developments in relations with the USSR. President Alfonsín visited Moscow in October to try to revive Soviet–Argentinian economic relations, and Soviet Foreign Minister Shevardnadze visited Mexico to prepare the ground for Gorbachev's planned visit to Mexico, Brazil and Argentina in late 1987. This would be the first trip to the area by a Soviet leader and would significantly expand the range of Gorbachev's initiatives around the world.

Chronologies: 1986

THE UNITED STATES AND CANADA

January

17 President Reagan issues 'secret finding' authorizing sale of US arms to Iran.

28 US space shuttle *Challenger* explodes soon after lift-off, killing all 7 crew members.

28– Leader of UNITA rebels in Angola, Jonas Savimbi, begins US visit to seek aid; talks with Secretary of State Shultz (29) and President Reagan (30).

February

5 Administration proposes Fiscal Year 1987 budget of $994 billion, including defence spending increase of 12% to $320.5 billion.

18 US announces military aid package for UNITA.

26 Canada suspends testing of US cruise missiles over its territory, after two missiles crash.

28 Packard Commission on defence management presents report to President Reagan.

March

7– US informs USSR that it must cut its mission to the UN from 275 to 170 by April 1988; USSR protests, denies spying charges (8).

13 House of Representatives rejects Administration's budget request (312–12).

20 After two-day meeting US President Reagan and Canadian prime minister Mulroney sign 5-year extension of North American Air Defense Agreement, due to expire in May.

April

18 *Titan* 34-D satellite launch vehicle explodes.

29 US and USSR resume commercial flights as agreed at November 1985 summit.

May

3 *Delta* rocket destroyed on lift-off and further satellite launches are suspended, leaving US with no satellite launch capability.

5 Chinese military delegation begins 16-day US visit.

7 Senate votes unanimously to undertake major reorganization of US military command structure based on Packard Commission report.

28 National Security Adviser McFarlane arrives in Tehran on secret visit for four days of negotiations on arms sales and hostages.

June

10 Report of commission investigating *Challenger* shuttle explosion blames failure of booster rocket joint and calls for complete overhaul of NASA's management and safety procedures.

20 Soviet air attaché expelled for spying.

July

7 Supreme Court rules automatic budget cuts mandated by Gramm/Rudman Bill unconstitutional.

17 Senate ratifies Anglo-American extradition treaty.

August

5 US and USSR sign cultural, scientific and educational exchange agreements.

15 President Reagan authorizes building of replacement for *Challenger* space shuttle.

23 Soviet UN employee Gennady Zhakarov arrested in New York for spying.

September

16– Filipino President Corazón Aquino begins 10-day US visit; meets President Reagan (17); addresses Joint Session of Congress (18); talks with World Bank to arrange loans (19).

17 US orders 25 Soviet UN diplomats to leave by 1 October.

19– Soviet Foreign Minister Shevardnadze meets President Reagan in Washington to discuss Daniloff affair (see USSR), delivers Gorbachev invitation to pre-summit meeting.

October

8 Chief State Department spokesman Bernard Kalb resigns in protest at August 'disinformation' campaign against Libya's Gaddafi by National Security Council.

15– Compromise budget adopted by House of Representatives and Senate (16) sets FY 1987 defence spending at $291.8 billion, cuts Administration's $5.3-billion SDI request to $3.5 billion.

19– USSR expels 5 US diplomats in response to US expulsion of Soviet diplomats; US expels 55 Soviet diplomats (21); USSR expels 5 more, withdraws Soviet workers from US missions in Moscow and Leningrad (22).

November

4 Iranian Parliament speaker Rafsanjani reveals former US National Security Adviser McFarlane visited Iran in May for negotiations on arms sales and release of US hostages in Lebanon.

4 Democrats gain control of Senate (55–45) in mid-term elections.

19– President Reagan assumes full responsibility for authorizing weapons shipments to Iran, denies link between deliveries and release of American hostages in Lebanon; Attorney General Meese reveals some money paid by Iran for US arms was diverted to Contra rebels (25); National Security Adviser Vice-Admiral Poindexter resigns and his subordinate, Lt-Col. North, is fired (25); Reagan appoints special commission under former Senator John Tower to investigate National Security Council and its role (26).

December

2– Frank Carlucci appointed US National Security Adviser, Reagan designates independent counsel to investigate Iran/Contra affair; Congress forms special committees to conduct full investigations (4).

30 US decides to impose 200% duties on some imports from EEC from 30 January 1987, after failure of negotiations on compensation for US grain sales lost due to Spain's entry into EEC.

THE SOVIET UNION

January

1 President Reagan allowed to broadcast New Year greeting on Soviet TV, General Secretary Gorbachev reciprocates on US TV.

February

18 Former Moscow Communist Party chief Viktor Grishin removed from Politburo, his successor in Moscow, Boris Yeltsin, appointed candidate member.

20 Soviet space station *Mir*, first stage of research and industrial complex, put into orbit.

25– Gorbachev opens 27th Communist Party Congress in Moscow with speech outlining foreign-policy priorities and plans for radical economic reform; head of defence industries Lev Zaikov made full Politburo member, and Nikolai Slyunkov and Yuri Soloviev made candidate members (6 March).

March

18 Soviet Union formally protests at entry of 2 US warships into territorial waters in Black Sea.

April

14– Soviet Foreign Minister Shevardnadze meets Chinese Deputy Foreign Minister Qian Qichen in Moscow, proposes Sino-Soviet summit meeting; China rejects proposal (16).

15 USSR cancels pre-summit Shevardnadze–Shultz meeting in Washington in protest at US air raid on Libya (see Middle East).

24 USSR and Britain sign co-operation agreement on energy conservation and development.

26– Major accident at Chernobyl nuclear plant near Kiev releases radioactive cloud over Europe; Swedish energy commission informs world, USSR evacuates 18½ mile radius around reactor (28); first Soviet official announcement says 2 killed (29).

May

14 Gorbachev breaks high-level silence on Chernobyl with frank and detailed discussion of accident.

23– Delegation of British MPs begins 10-day USSR visit; Gorbachev proposes bilateral arms-control negotiations (26).

June

10– Warsaw Pact leaders begin two-day meeting in Budapest; issue statement on conventional force reductions in Europe (11) (see also Arms Control).

July

7– French President Mitterrand follows US visit with 4-day trip to Moscow and 16 hours of discussions with Gorbachev.

20– West German Foreign Minister Genscher begins three-day talks concentrating on arms control with Soviet leaders in Moscow; signs science and technology co-operation accords (22).

28 In major speech in Vladivostok aimed at improving relations with Asia, Gorbachev announces withdrawal of 6 Soviet regiments from Afghanistan, proposes removal of Soviet troops from Mongolia.

31 In Khabarovsk, Gorbachev calls for major economic reforms.

August

7 Former CIA official Edward Howard, accused of spying in US and missing since September 1985, granted asylum in USSR.

21 Soviet report on Chernobyl for International Atomic Energy Agency blames human error, puts death toll at 31 and promises improved safety measures.

30 US journalist Nicholas Daniloff arrested in Moscow on spying charges.

September

10 Observers from 9 NATO countries travel to Czechoslovakia to monitor 3-day military exercises.

27 Western journalists visit main Soviet nuclear weapons test site near Semipalatinsk for first time.

29– US and USSR announce Reagan and Gorbachev will meet in Reykjavik in October; USSR allows US journalist Daniloff to leave, announces it will free dissident Yuri Orlov; Gennady Zakharov allowed to leave US (30).

October

2 Soviet Space Institute delegates sign protocol with British National Space Centre scientists providing for joint space research.

3– Fire on Soviet *Yankee*-class nuclear submarine on patrol in North Atlantic kills 3; submarine sinks under tow some 700 miles north of Bermuda (6).

November

3 COMECON leaders begin two-day meeting in Bucharest.

19 Supreme Soviet adopts law legalizing 29 types of private enterprise in USSR.

December

16– Gennady Kolbin replaces Politburo member Dinmukhamed Kunayev as Party leader in republic of Kazakhstan; change provokes two days of riots in the capital, Alma Ata (17–19).

19 Dissidents Andrei Sakharov and wife Yelena Bonner released from internal exile and return to Moscow (23).

EUROPE

January

1 Spain and Portugal join European Economic Community (EEC).

7 Greece signs General Security of Military Information Agreement with US, allowing purchase of F-16 fighters.

9 British defence minister Heseltine resigns over government policy on bids for Westland helicopter company; George Younger appointed successor.

16 West German Chancellor Kohl and French President Mitterrand meet in Baden-Baden, agree to increase military co-operation.

20 In Lille Mitterrand and British premier Thatcher announce contract for building of Channel Tunnel.

24 Northern Ireland by-elections in 15 seats resigned by Unionists in protest at October 1985 Anglo-Irish accord: 1 Social Democratic/Labour and 14 Unionist MPs elected.

February

3– France expels 4 Soviet diplomats for spying; USSR retaliates by expelling 4 French diplomats (4).

5– Italy expels Soviet diplomat and Aeroflot official; USSR expels Italian diplomat and industrialist (6).

6 France announces increase in internal security after terrorist bomb attacks in Paris leave 21 injured.

6 Britain and USSR sign 5-year economic and industrial co-operation agreement.

11 USSR releases dissident Anatoly Shcharansky and three others in exchange for 5 Eastern-bloc spies.

12 Britain and France sign Channel fixed-link treaty at Canterbury ceremony.

12 Westland helicopter company shareholders vote to accept US/Italian rescue plan by Sikorsky and Fiat.

16 Former Portuguese prime minister Mario Soares wins Presidential elections with 51.2% in second round of voting.

16 4 Argentinian parliamentarians begin 7-day visit to Britain – first since Falklands conflict.

17– Nine of twelve EEC members sign 'Single European Act' reform package; Italy, Greece and Denmark, who boycotted ceremony, sign later (28).

28 At Franco-German Paris summit Mitterrand announces France will consult West Germany before using 'pre-strategic' weapons on its territory.

28 Swedish prime minister Olaf Palme assassinated in Stockholm.

March

3 Northern Ireland Unionists call 24-hour strike, accompanied by widespread violence, in protest at Anglo-Irish agreement.

5- Avalanche kills 16 Norwegian soldiers preparing for 20,000-man NATO Arctic exercises; exercises cancelled (6).

7 Greek parliament approves revisions of Constitution to limit presidential powers.

9 Mario Soares sworn in as Portugal's first civilian president in 60 years.

12 In Spanish referendum 52.5% support NATO membership.

12 In Florence British prime minister Thatcher and Italian prime minister Craxi discuss terrorism, sign first British-Italian extradition treaty in 100 years.

16– In French general elections right-wing RPR/UDF coalition win 286 of 577 seats in *Assemblée Nationale*; RPR leader Jacques Chirac appointed prime minister (20).

16 Switzerland votes 3–1 against joining United Nations.

24 Czech Communist Party begins 5-day congress in Prague.

27 West Germany and US sign Memorandum of Understanding on German companies' participation in SDI research.

April

2 Bomb kills 4 passengers on TWA flight approaching Athens airport.

2 Bulgarian Communist Party begins 4-day congress, Party leader Zhivkov calls for major changes in economic management.

5 Bomb explosion in West Berlin discotheque kills US soldier and Turkish woman, injures more than 200.

17– East German Communist Party Congress begins in East Berlin; addressed by Soviet General Secretary Gorbachev (18, see Arms Control); 4 new Politburo members elected, including new Defence Minister Kessler (21).

17 Syrian attempt to blow up El Al flight from London to Tel Aviv foiled by security guards before take-off.

21 Turkish-Cypriot government of Northern Cyprus accepts UN draft plan for island reunification, Greek-Cypriot government presents its own counter-proposals.

25 After 2-day meeting in the Hague, EEC ministerial group on crime and terrorism (Trevi Group) agrees to co-operate with US and Council of Europe members to combat terrorism.

28 In Madrid, West European NATO members agree to co-operative production of military aircraft.

May

2 Norwegian 3-party coalition under Kare Willoch resigns, Labour Party leader Mrs Gro Harlem Brundtland agrees to form government.

4 First-round voting in Austrian Presidential elections fails to give former UN General Secretary Kurt Waldheim or rival candidate Kurt Steyrer required 50% of votes.

6 East and West Germany sign cultural, scientific and sports accord after 13 years of negotiations.

9 Spain expels Libyan Consul-General for conspiring against the government.

10– Britain expels 3 Syrian diplomats for complicity in terrorism; Syria expels 3 British diplomats (11; see also Middle East 24 October).

22 NATO defence ministers, meeting at Defence Planning Committee in Brussels, approve US force goals for chemical weapons modernization.

29 The IMF approves Poland's application to rejoin after 36-year absence.

June

4 Belgian House of Representatives bans deployment of chemical weapons on Belgian territory.

8– Kurt Waldheim elected President of Austria in second-round voting; Chancellor Sinowatz, Foreign Minister Gratz and 2 other ministers resign (9, 10); Franz Vranitzky sworn in as new Chancellor (16).

11 British SDP/Liberal Alliance joint commission on defence calls for cancellation of *Trident* programme but fails to decide whether *Polaris* should be replaced.

22 Spanish elections return González government to power.

24 Britain signs contracts (worth $14.3 million) with US for SDI research.

25 Yugoslav Communist Party begins 5-day congress in Belgrade.

26– Premier Craxi's Italian government loses vote on local government finance bill; government resigns (27).

29– Polish Communist Party congress opens in Warsaw; addressed by Soviet leader Gorbachev (30).

29 European Space Agency agrees to support French-designed *Hermès* manned spacecraft programme.

30 *Eureka* ministerial meeting in London approves 62 new development programmes, agrees to establish secretariat in Brussels, admits Iceland as 19th member.

July

2 Turkish premier Ozal begins 3-day visit to Republic of Northern Cyprus.

3 Polish defence minister Gen. Siwicki and foreign minister Orzechowski appointed to new Politburo with five others.

10 French government decides against production of neutron bomb.

13– Soviet foreign minister Shevardnadze begins 3-day visit to Britain; signs 3 agreements, including long-term programme for economic collaboration.

17 Polish government offers amnesty to nearly all political prisoners and up to 20,000 criminals if they renounce crime.

August

1 After prolonged negotiations, Craxi forms new Italian government.

27 West German Social Democratic Party (SPD) conference in Nuremberg votes to remove all US cruise and *Pershing* nuclear missiles from West Germany and phase out nuclear power.

September

8– 'Solidarity Committee of Arab Political Prisoners' begins 10-day wave of bomb attacks in Paris; government announces new anti-terrorist measures, including visa requirement for all foreigners except EEC and Swiss nationals (14).

11 Polish government announces it will release all political prisoners by 15 September.

13 Austrian government of Chancellor Vranitzky collapses, general elections called for November.

19 Italy and US sign agreement on participation of Italian companies in SDI research.

25 Trevi Group meeting in London agrees to improve European counter-terrorism co-operation.

26 Special session of International Atomic Energy Agency in Vienna adopts conventions to provide early notification of nuclear accidents and international assistance during emergencies.

30– In Poland, Solidarity announces formation of public 'Temporary Council' to campaign for trade union freedom; government bans new council (4 October).

October

2 British Labour Party conference votes to remove all nuclear weapons from Britain and renegotiate terms of US military base agreements.

19 In joint declaration to mark 30th anniversary of Hungarian Revolution, East German, Hungarian, Polish and Czech dissidents call for political democracy in Eastern Europe.

24 Nezar Hindawi sentenced by UK court to 45 years' imprisonment for El Al bombing attempt at Heathrow (see 17 April).

24 West German Chancellor Kohl compares Soviet public relations techniques to those of Nazi propaganda chief Goebbels in *Newsweek* interview, chilling German–Soviet relations.

26 Prime ministers González of Spain and Cavaco Silva of Portugal begin 2-day summit in Guimaraes, Portugal, agree to increase economic co-operation and improve bilateral relations.

November

3 Albanian Communist Party begins first congress since death of Enver Hoxha.

5 French government approves Ffr 474-billion 5-year defence plan, including major modernization of nuclear forces and development of chemical weapons.

10 Greece and US sign defence industrial co-operation agreement stipulated in 1983 bases accord.

21– Romanian President Ceausescu announces large cuts in armed forces; referendum on unilateral 5% cut in military spending passed almost unanimously (24).

23 Austrian general elections deny Socialist Party a parliamentary majority and double right-wing Freedom Party's vote, leading to negotiations for new government.

27– French students begin two weeks of demonstrations in protest at government's education reform bill; one student killed in ensuing violence (6 Dec.); government drops bill (8 Dec.).

December

5 Strasbourg meeting of European Council Justice and Interior ministers extends Trevi Group anti-terrorism information network to all 21 members of Council.

9 Trevi Group meeting in London produces document analysing terrorist threat in Europe, Greece refuses to sign.

12 Turkey and US agree on terms of new Defence and Economic Co-operation Agreement after 15-month negotiations.

18 British government cancels 9-year *Nimrod* programme and says it will buy 6 Boeing AWACS early-warning aircraft.

THE MIDDLE EAST AND NORTH AFRICA

January

7 US introduces economic sanctions against Libya and asks all Americans to leave.

13 Israeli Cabinet accepts principle of arbitration in Taba dispute with Egypt, should bilateral conciliation talks fail.

13– Failed assassination of South Yemen President Nasser Mohammed leads to fierce fighting between rival Marxist factions in Aden; French, Soviet and British ships begin evacuating foreign nationals (17); premier Haidar Abu Bakr al-Attas appointed interim President by victorious radical faction (24).

17 Israel and Spain establish formal diplomatic relations.

February

2 US Administration sets aside planned $1.9-billion arms sale to Jordan as result of Congressional opposition.

2– Soviet First Deputy Foreign Minister Kornienko visits Iran for 2 days, concludes agreement to resume Aeroflot flights to Tehran and improve economic ties.

4 Israeli fighters force Libyan aircraft, believed to be carrying Palestinian guerrilla leaders, to land in Israel.

8 Haider Abu Bakr al-Attas named President of South Yemen.

9– Iranian troops cross Shatt al-Arab and seize the Faw peninsula; Iraq counter-attacks (14).

19 Jordan's King Hussein announces end of political co-ordination with PLO and abandonment of his Middle East peace initiative.

25– Egyptian Central Security Force conscripts riot; Army ends riots and President Mubarak dismisses Interior Minister Rushdi (28).

March

2– Nablus Mayor Zafr al-Masri shot dead by Palestinian extremists; moderate Palestinian nominees for other mayoral posts on the West Bank begin to withdraw their candidacies (3).

5– In Lebanon *al-Jihad al-Islami* claims to have killed French researcher Michel Seurat in reprisal for France's deportation of 2 pro-Iranian Iraqi students to Iraq; seizes four-man French TV crew in Beirut (8).

19 Terrorist group 'Egypt's Revolution' claims responsibility for killing Israeli embassy member and injuring 3 others in Cairo.

21 UN condemns Iraq for using chemical weapons against Iran.

24– Libyan missiles fired at US Navy fighters over Gulf of Sirte; US bombs Libyan missile base and patrol boats in retaliation (24–25); US ends manoeuvres in area (27).

April

1 France announces withdrawal of its 45-man military observer force from Beirut.

5– In Saudi Arabia, US Vice-President Bush fails to resolve policy differences on oil pricing at the start of 10-day trip that also takes in Bahrain (7), Oman (9) and North Yemen (10–12).

15– US aircraft from Sixth Fleet carriers and British bases bomb targets in Libyan cities of Tripoli and Benghazi in retaliation for alleged Libyan complicity in 5 April bombing of West Berlin discotheque; 2 kidnapped Britons and an American murdered in Beirut in reprisal (17); foreign nationals begin evacuating from Muslim West Beirut (20); EEC foreign ministers meeting in Luxembourg agree to restrict number and movement of Libyan diplomats in Europe (21); US, Britain and France veto UN Security Council draft resolution condemning US bombing of Libya (22).

26– Tension between Qatar and Bahrain flares over building of Bahraini coastguard station on disputed border reef; Saudi Arabian defence minister arrives to mediate (27); crisis eventually resolved by simultaneous troop withdrawals from the area (15 June).

May

5 Syrian President Assad begins two-day visit to Jordan to discuss improvements in bilateral relations with King Hussein.

6 Israel's Defence Minister Rabin signs an accord in Washington stipulating guidelines for Israeli participation in SDI research.
7– US Congress bans President Reagan's proposed $354-million arms sale to Saudi Arabia; Saudi Arabia drops request for 800 *Stinger* anti-aircraft missiles (20); Reagan vetoes congressional ban (21).
12 Libya expels 36 EEC diplomats in retaliation for European reductions of Libyan People's Bureaux staff.
24 Mrs Thatcher makes first-ever visit to Israel by British premier.
24– King Hussein visits Syria and Iraq (26) to try to mobilize support for resolution of Iran–Iraq war.
26– Deputy to Libya's Gaddafi begins four-day visit to Moscow; talks with General Secretary Gorbachev and Defence Minister Sokolov (27); Libya announces Soviet agreement to provide arms (27).

June

5 US Senate fails to override President's veto of its ban on Saudi Arabia arms sale.
11 In Washington King Hussein declares 30-year special relationship ended because of US Congressional refusal to sell arms to Jordan.
16 George Saadeh replaces Elie Karameh as leader of Lebanese Christian Phalange Party.
20 2 members of kidnapped French TV crew released in response to recent changes in French Middle East policy.
21 Spanish riot police sent to enclave of Melilla after 2 days of clashes between right-wing Spaniards and Muslims.
24 Lebanese Army troops deployed around Palestinian camps near Beirut to enforce Syrian-mediated truce between Palestinians and Shi'ite Muslim militia after a month of fighting.
25 Israeli Shin Bet secret service chief Avraham Shalom and others resign over reported cover-up of deaths of 2 Arabs taken from hijacked bus in Gaza strip in 1984.

July

1– Kuwaiti cabinet resigns after dispute with National Assembly; Emir dissolves democratically elected Assembly, asks premier Saad al-Abdullah al-Saban to form new government (3).
1– Presidents of North and South Yemen meet for 2 days in Libya and pledge to work towards unity.
8– Jordanian government closes all 25 offices of *Fatah* mainstream PLO; King Hussein announces $1.3-billion economic development plan for West Bank (12).
10– Retaliating for Palestinian guerrilla raid which killed two soldiers in Israel, Israeli aircraft attack Palestinian camps in southern Lebanon; hit camps outside Beirut (14).
21 Justice minister Modai resigns over prime minister Peres' handling of the Shin Bet scandal.
22 Peres begins 2-day talks with Morocco's King Hassan, Syria breaks diplomatic relations with Morocco in protest.
27 US hostage Father Lawrence Jenco released in Lebanon.
27 King Hassan of Morocco resigns as Chairman of Arab League summits after Arab criticism of his meeting with Israeli premier Peres.
27– US Vice-President Bush arrives in Israel at the start of 10-day Middle East tour; also goes to Jordan (30), Egypt (3 August).

August

7– Israeli and Egyptian negotiators agree terms of arbitration over disputed Taba enclave in the Gulf of Aqaba; Israeli cabinet unanimously approves terms (13).
10 Israel begins a series of attacks on alleged Palestinian guerrilla bases in southern Lebanon.
10 Fighting erupts between rival Christian militia groups in Beirut, as Phalangists try to topple anti-Syrian Lebanese Forces leader, Samir Geagea.
12 Iraqi aircraft attack Iran's Sirri Island in southernmost area of the Gulf.
14 Spain grants PLO office in Madrid semi-diplomatic status.
18 Israeli and Soviet representatives meet in Helsinki – first official contact between the 2 countries since 1967.
24 Egypt and US begin 4-day joint naval exercise in Mediterranean.
29 Morocco abrogates 1984 unity pact with Libya after Libyan criticism of King Hassan's July meeting with Israeli premier Peres.
30– Iraqi Parliament speaker Saadoun Hamadi announces Iraq is willing to accept internationally guaranteed non-aggression pact to end war with Iran; Iran begins new offensive in northern Iraq (1 September).

September

2 Lebanese cabinet meets in Beirut for first time in 10 months and agrees to new cease-fire.
4– 3 French soldiers with UNIFIL peace-keeping force in Lebanon killed in bomb attack; French contingent deployed to safer positions near main base at Nakoura (21).
10– Egyptian President Mubarak and Israeli premier Peres agree composition of Taba arbitration team; begin two-day summit in Alexandria (11).
22 Peres meets Soviet foreign minister Shevardnadze at UN to discuss restoration of diplomatic relations.
23 Egypt elevates its *chargé d'affaires* in Tel Aviv to the status of ambassador.
23 Israel refuses to comply with a UN Security Council resolution calling for complete withdrawal of Israeli troops from southern Lebanon.

October

7– Israeli premier Peres reveals Israel has been talking indirectly with Jordan via the US and has agreed to an international conference; resigns his post in accordance with the 1984 power-sharing agreement (10); *Likud* leader Yitzhak Shamir becomes prime minister (20).
12 Medhi Hashemi, member of staff of Khomeini's designated successor Ayatollah Montazeri arrested on suspicion of sedition.
15– Grenade attack on military ceremony at Wailing Wall in Jerusalem kills 1, injures 69; Israeli aircraft bomb Palestinian camp near Lebanese port of Sidon, 1 airman captured by *Amal* militia (16).
24– Britain breaks relations with Syria after London court finds Syrian involvement in April attempt to blow up El Al jet (see Europe); Syria closes its air and sea space to British craft and expels British diplomats; Britain fails to persuade EEC to back sanctions (27).
29 Saudi Arabian oil minister Sheikh Yamani replaced after 24 years in office.

November

2 US hostage David Jacobsen released in Beirut.

9 Egypt's President Mubarak chooses new premier, Alef Sedki, to tackle severe economic problems.

10– EEC nations, except Greece, approve measures against Syria which include ban on arms sales; US announces sanctions (14).

11 2 French hostages, kidnapped in Lebanon, released in Syria after France agrees to repay part of $1-billion loan to Iran.

16– Iranian-backed Kurdish guerrillas launch 3-pronged offensive against northern Iraqi oilfield; Iraq begins counter-attack (17); Iraq bombs Iran's newest oil terminal on Larak Island at mouth of Gulf (25).

26 France confirms it is cutting its UNIFIL contingent from 1,380 to about 500, other nations agree to increase their troops to compensate.

27– West Germany expels 4 Syrian diplomats, appoints no new ambassador to Damascus and ends development aid in protest at Syrian involvement in March bomb attack in Berlin; Syria orders 3 West German envoys out of Damascus (28).

December

12 Iran and USSR sign economic co-operation protocol.

16 Israel allows Abu-Zaim's pro-Jordanian rebel PLO faction to open office in Hebron on occupied West Bank.

23– Iran and Iraq escalate air raids on military and civilian targets; Iran begins offensive on Shatt al-Arab waterway (25).

24 Another of the kidnapped French TV crew released in Lebanon; French premier Chirac thanks Algeria, Syria, Lebanon and 'certain Palestinians' for their help.

25– Hijacked Iraqi airliner, en route from Baghdad to Amman, crash lands, killing 62.

SUB-SAHARAN AFRICA

January

1– South Africa, accusing Lesotho of harbouring ANC guerrillas, restricts movement across border; Lesotho premier Chief Jonathan overthrown and replaced by military regime under Maj.-Gen. Justin Lekhanya (20); King Moshoeshoe II granted executive and legislative powers by new leaders (22); border controls lifted after new government deports ANC supporters to Zambia (26).

6 Gen. Samuel Doe sworn in as President of Liberia, new constitution takes effect.

16– Iran's President Khamenei begins tour of Front Line African states in Tanzania, talks with government, ANC and SWAPO representatives; visits Mozambique (18), Angola (19) and Zimbabwe (20–22).

19 Burkina Faso and Mali agree to withdraw troops from disputed border area.

22– At least 30 killed in tribal clashes near Durban, South Africa.

26– Ugandan National Resistance Army takes Kampala, dissolves ruling

Military Council led by Gen. Okello; Yoweri Museveni sworn in as President (29); names new cabinet, including himself as defence minister and Dr Sam Kisekka as premier (30).

30– 9-member Southern Africa Development Co-ordination Conference meets in Harare, reviews progress in disengaging their economies from South Africa.

31 South African President P.W. Botha offers black leaders advisory role in the legislative process.

February

11 Nigeria and Britain agree to resume full diplomatic ties.

11– In Chad, Libyan-backed rebels begin an offensive to the south of 16th parallel.

12– Ivory Coast restores diplomatic ties with Israel; and with the Soviet Union (20).

15– Violence escalates, riots kill at least 14 in black townships in South Africa.

16– French aircraft bomb northern Chad airfield used by rebels; Ndjamena airport attacked in retaliation (17).

16 Commonwealth Eminent Persons Group, commissioned by November 1985 Commonwealth Summit to find ways to promote discussion of reform in South Africa, arrives in Johannesburg to begin mission.

March

1– Nigeria reopens all land borders (except with Chad), closed since April 1984; 10 army officers, convicted of December 1985 coup plot, are executed (5).

4– President Botha proposes UN Resolution 435 for settling Namibian question be implemented from 1 August, if Cuban forces are withdrawn from Angola by then; Angola rejects this (9); Angola asks UN to resume negotiating effort over Namibia, claiming US, having given UNITA military aid, cannot now be credible mediator (18).

7 State of Emergency lifted throughout South Africa and most of those arrested under it released; police fire on black schoolchildren in Eastern Transvaal, killing 2 and injuring over 80.

13 US says it will send Chad $10 million emergency military aid.

18 Botswana closes ANC's Gaborone office and expels all its representatives after informal agreement with South Africa.

29 All Mozambique ministries put under authority of 3 Politburo members, Gen. Chipande to handle war against RENAMO; President Machel leaves for 5-day visit to USSR.

April

2– Bishop Tutu calls for punitive economic sanctions against South Africa, order banning Winnie Mandela from her home in Soweto lifted; South Africa announces plans to abolish pass law system but at the same time also increases Law and Order Ministry's emergency powers (23).

3– Kwazulu–Natal Indaba, bringing together black and white political leaders, begins talks aimed at setting up multi-racial provincial legislative assembly in Natal.

26 Sudan's newly-elected civilian parliament meets but adjourns to let major parties continue negotiations on the formation of a coalition government.

May

1 Over 1 million South African blacks boycott work and school for a day; 2 killed, hundreds injured as police and army break up large demonstration in Transvaal protesting at plans to make Kwandebele fifth independent tribal homeland (14); at least 26 die in 4 days' fighting between conservative and radical blacks in Crossroads squatter camp (17); government unveils plans for multi-racial advisory 'National Council' and expands police powers of detention (22).

6 Parties in Sudan agree on division of power in cabinet, government under Sadeq al-Mahdi takes over from transitional military council.

7 Talks between Ethiopia and Somalia on Ogaden border dispute result in little progress.

14– Indian premier Gandhi begins tour of Front Line states in Zambia; goes to Zimbabwe (15), Angola (16) and Tanzania (17).

19– South African forces raid alleged ANC guerrilla bases in Botswana, Zambia and Zimbabwe, Eminent Persons Group withdraws from South Africa in protest; US expels South African military attaché (23); Denmark announces ban on trade with South Africa from 1 December (30).

22 South African foreign minister Pik Botha abandons address to National Party rally when extreme-right Afrikaner Weerstandsbeweging supporters take over meeting.

23 President Barre of Somalia seriously injured in car crash in Mogadishu.

June

9– In renewed clashes between rival black groups in South Africa's squatter camp at Crossroads, at least 20 people are killed, hundreds wounded and an estimated 20,000 are left homeless; Commonwealth Eminent Persons Group publishes report on its mission, calls for imposition of economic sanctions – South Africa declares nationwide State of Emergency (12); President Botha and Bishop Tutu meet for the first time in six years to discuss the situation (13); car bomb explosion in Durban kills 3, injures 69 (14); anniversary of 1976 Soweto uprising marked by mass boycott by black workers (16).

16– 5-day World Conference on Sanctions against South Africa, (organized by UN Special Committee against Apartheid, Organization of African Unity (OAU) and Non-Aligned Movement) meets in Paris; calls for mandatory, comprehensive economic sanctions (20); Hague meeting of EEC heads of state delays decision on sanctions issue for 3 months and commissions British foreign minister Howe to go to southern Africa to try for a dialogue (27).

20 South Africa says it will not implement UN Resolution 435 on Namibian independence on 1 August, because Cuban troops are still in Angola.

July

1– South African pass laws are repealed, uniform identity cards are introduced for all citizens; all restrictions on Mrs Winnie Mandela are lifted (7).

8 Only 1 month after appointing him, Tunisian President Bourguiba replaces his Prime Minister and constitutional successor Mzali with finance minister Sfar.

9– Sir Geoffrey Howe begins EEC mission to southern Africa in Zambia, ANC refuse to meet him but he talks with President Kaunda; has further discussions in Zimbabwe (10) and Mozambique (11); meets South African President Botha (23); faces strong attack on British and US policy from Kaunda (24); returns to South Africa (25); talks with President and Foreign Minister with no result (29).

10 US says it will withhold $13.5 million aid to Zimbabwe till apology is received for Zimbabwean minister's anti-American remarks at 4 July reception in Harare.

21 Guinea Bissau government executes 6 convicted coup plotters, including former Vice-President Paulo Correia.

28– 3-day OAU summit in Ethiopia calls for voluntary sanctions against Britain because of opposition to sanctions against South Africa.

30 US ambassador to Zambia meets 3 ANC leaders in Lusaka in first official US contact with the movement.

August

3– Commonwealth meeting in London fails to agree on sanctions against South Africa, Britain holding out against other six nations; Southern African Development Co-ordination Conference (SADCC) and Front Line states hold summit in Zambia but also fail to agree on sanctions (21).

4 Guerrillas cut Mozambique oil pipeline from Zimbabwe to port of Beira.

12– South African troops attack government forces in southern Angola but fail to seize base at Cuito Cuanavale.

16 Sudanese SPLA rebels shoot down civilian aircraft, killing all 57 passengers, all flights in the area (including famine relief) are suspended.

18– South Africa announces it is holding 8,501 under the state-of-emergency regulations; violent clashes in Soweto between police and blacks protesting at evictions of those taking part in rent boycotts – 21 killed, almost 100 injured (26).

25– Israeli premier Peres on 2-day official visit to Cameroon; full diplomatic relations restored (26).

September

1– Week-long 5th conference of Non-Aligned Movement dominated by sanctions against South Africa; in Brussels, EEC foreign ministers adopt South African sanctions package weaker than that agreed by heads of government in June (16); Japan adopts same measures plus visa ban for South African visitors (19); Reagan vetoes bill of tough economic sanctions against South Africa (26); veto overridden in House of Representatives (29) and Senate (2 October).

7 Desmond Tutu enthroned as Anglican Archbishop of Southern Africa in Cape Town.

24– Troops crush coup attempt in Togo; government claims rebels came from Ghana, closes border, requests military aid from France (25).

October

3– 2 Ugandan cabinet ministers and former Vice-President arrested; accused of plotting to overthrow Museveni government (7).

6 After land-mine explosion injures 6 on South Africa–Mozambique

border, South Africa says Mozambique allows ANC terrorists to operate from its territory, imposes ban on recruiting Mozambique workers.

8 Somalia and USSR restore diplomatic relations, broken in 1977.

19 President Machel of Mozambique and several cabinet ministers killed in air crash just inside South Africa–Mozambique border; South Africa denies complicity, invites Mozambique and USSR to join in crash inquiry (21).

20– General Motors announces withdrawal from South Africa (first of many large foreign companies); Red Cross Conference in Geneva votes to exclude South African government delegates from its proceedings (25); Pretoria orders all Red Cross representatives to leave South Africa (26).

23 Deposed Central African Emperor Bokassa, returning from exile in France, arrested on arrival.

26– Mugabe announces unlimited support for Mozambique against RENAMO guerrillas; RENAMO declares war on Zimbabwe and threatens attacks on Zimbabwe targets (28).

27 Ethiopian foreign minister Goshu Wolde resigns while visiting US, protesting at government's authoritarian policies.

30 Goukouni Oueddei, leader of Chad rebels, shot and then arrested together with 55 others, in Libyan capital.

November

3 Mozambique's foreign minister Joaquim Chissano chosen to succeed Machel as President; sworn in (6).

6 South Africa claims to have found at Machel air crash site documents revealing Mozambique and Zimbabwe were plotting to overthrow Banda's Malawi government.

26 South Africa reverses expulsion order against Red Cross officials and allows organization to continue work in the country.

28 Kwazulu–Natal Indaba ends with 24–2 vote for joint black/white legislature for the province, with universal suffrage; Pretoria rejects proposals (30).

December

2 Kenyan parliament votes for constitutional amendment to increase President's powers.

8– 100% rise in maize meal price sparks riots in several towns in the north Zambian copperbelt and Lusaka; land borders closed to outgoing traffic, dawn-to-dusk curfew imposed (9); troops used to curb riots (10); price rise abandoned (11).

11– South Africa introduces new press restrictions; President Botha claims reason is need to block planned ANC Christmas bombing campaign (12).

11– In Chad, Libyan troops launch offensive against former rebel allies; President Habré accuses Libya of using napalm (12); 2 French aircraft drop weapons and food for rebels in the Tibesti mountains (17); US announces it will meet Habré's aid request with $15 million worth of military equipment (18); Chadian forces begin a counter-offensive (18).

22 World Court rules on territorial dispute between Mali and Burkina Faso, dividing disputed area between them.

ASIA AND AUSTRALASIA

January

7 Mauritian Foreign Minister and three other ministers resign in scandal after Mauritian MPs arrested in Netherlands for drug smuggling.

10– Trust territory Republic of Palau signs 'Compact of Free Association' with US, granting military concessions in return for self-rule and economic aid; islanders vote 3–1 for pact (23 Feb.).

15– Soviet Foreign Minister Shevardnadze begins 4-day visit to Japan; discusses peace treaty with Foreign Minister Abe (15–16) but dispute over Northern Territories unresolved.

20 North Korea suspends talks with South Korea in protest at US/South Korean joint military exercises.

22– 3 Sikhs, including two police bodyguards, sentenced to death for 1984 assassination of Indian Prime Minister Indira Gandhi; Sikh extremists in Punjab take over Golden Temple in Amritsar (26).

February

15– Ferdinand Marcos declared winner of Philippines' presidential election but opposition and independent observers claim widespread fraud, Marcos dismisses Head of the Armed Forces Gen. Ver; US President Reagan sends special envoy Philip Habib to Manila for talks on reform (16); amid mass demonstrations opposition candidate Corazon Aquino, Defence Minister Enrile and new Head of the Armed Forces Ramos take over two military camps in Manila (22); Marcos cedes power and flies to Hawaii, Mrs Aquino sworn in as president (25).

27 Malaysian Deputy Prime Minister Datuk Musa Hitam resigns, causing government crisis.

March

11– Dissident Kim Young Sam and New Korea Democratic Party leader Lee Min Woo lead thousands of demonstrators on march through Seoul calling for constitutional revision; National Liaison Organization for Democratization launched (17).

21 Premier Zhao Ziyang announces China will conduct no more atmospheric nuclear tests.

25 Japanese left-wing terrorists fire 3 rocket bombs at Emperor's Palace and 3 at US Embassy in Tokyo.

25 Filipino President Aquino abolishes constitution and National Assembly, sets up Provisional Government with interim constitution.

28– Indian Prime Minister Gandhi sends Congress (I) Vice-President Arjun Singh to Punjab to handle extremist violence that has killed more than 50; Punjab state police chief replaced (29); Punjab governor replaced (1 April).

April

7– Chun Doo Hwan in Britain on first European visit by S. Korean head of state; visits West Germany (10), France (13) and Brussels (18).

9 Australia rules out involvement in US SDI research.

10 Pakistan opposition leader Benazir Bhutto returns from exile and begins nationwide political tour.

10 Sydney newspaper article, accusing Indonesian President Suharto and his family of corruption, sparks diplomatic row between Australia and Indonesia.

22 Soviet and Afghan troops in south-east Afghanistan destroy key rebel stronghold at Zhawar.

29– Sikh extremists in Golden Temple at Amritsar proclaim independent state of Khalistan; police storm temple, seize about 300 extremists (30).

29– In Sri Lanka at least 200 killed in fighting between rival Tamil guerrilla groups.

May

1 Reagan has talks with ASEAN leaders in Bali *en route* to Tokyo economic summit.

1 Coalition government of Thai premier Gen. Prem Tinsulanonda defeated in vote on economic policy, King Bhumipol dissolves parliament and calls early elections.

3 Students riot in Inchon, South Korea.

4– Afghan Party leader Babrak Karmal resigns, replaced by security chief Najibullah; Karmal remains President of the Revolutionary Council and becomes member of collective leadership with Najibullah and Prime Minister Sultan Ali Keshtmand (15).

4– Seven-nation economic summit begins in Tokyo; leaders agree to increase international economic co-operation, combat terrorism, and increase nuclear energy safety (6).

7 General election held in Bangladesh amid violence; pro-government Jatiya Party declared winner but cannot muster two-thirds majority needed for constitutional changes (20).

7 US Secretary of State Shultz begins two-day visit to South Korea, expresses strong support for President Chun Doo Hwan.

17– Sri Lankan government begins major offensive to oust Tamil guerrillas from strongholds in Jaffna peninsula; India condemns offensive for hampering its peace efforts (19).

17– In Hong Kong representatives of China and Taiwan meet for first time for talks on returning Taiwanese airliner (flown to China by defector on 3 May); talks end successfully (19).

28 Indonesia and Papua New Guinea end talks on friendship and co-operation treaty with agreement on main principles.

29 Leaders of South Korean ruling and opposition parties agree to establish National Assembly Committee to draft constitutional reforms.

June

3 Australian defence review (Dibb Report) calls for restructuring armed forces for strategy of self-reliance.

8– General Secretary of Chinese Communist Party Hu Yaobang begins 4-day visit to Britain to discuss Hong Kong's future; travels to West Germany (12), France (17), and Italy (19).

18 Sri Lankan government approves proposals to settle Tamil conflict by devolving power to new provincial councils.

19– New Zealand and France agree to UN arbitration in dispute over 1985 sinking of Greenpeace ship *Rainbow Warrior*, France ends ban on New Zealand imports; after mediation France agrees to apologize formally and pay $7 million compensation in return for release of two French agents to 3-year confinement on Pacific atoll of Hao (7 July).

22 Vietnamese government dismisses 8 ministers, including Politburo member To Huu, after economic reforms fail.

24 ASEAN Foreign Ministers' annual meeting in Manila calls for working group to draft South-east Asian nuclear-free zone treaty.

25 Rebel leaders from Mizoram, eastern India, sign agreement with Delhi ending 25-year revolt and giving Mizoram full statehood.

30 USSR establishes diplomatic ties with Vanuatu.

July

6 Premier Nakasone's Liberal Democratic Party wins by landslide in Japanese general elections.

8 Failed coup attempt in Philippines, led by ex-President Marcos' foreign minister Arturo Tolentino.

10– Vietnamese Communist Party General Secretary Le Duan dies; Head of State Council, Truong Chinh, replaces him (14).

13 Moderate Tamil leaders and Sri Lankan government begin first round of peace talks in Colombo.

15– Pakistani prime minister Junejo and foreign minister Khan begin US visit; sign agreement to buy advanced US technology (17).

15– India accuses China of June incursions into Arunachal Pradesh; China denies charges (16).

27– General elections in Thailand; Gen. Tinsulanonda appointed to third term as prime minister (5 August)

August

1 Seven-day round of Pakistan/Afghanistan proximity talks begins.

3 Snap Malaysian general elections return premier Mahathir's National Front Party to power.

5 Peace talks begin between Filipino government and Communist rebels.

8 Japanese education minister Fujio sacked for asserting in interview that 1910 Japanese invasion of Korean peninsula was undertaken with Korean consent.

9 Chinese Vice-Foreign Minister Lin Shuqing ends visit to Mongolia by signing agreement on consular relations.

11 US Secretary of State Shultz announces US is suspending its security obligations to New Zealand stipulated by the ANZUS Treaty.

12 First meeting of South Asian Association of Regional Co-operation (SAARC) Council of Ministers agrees on Kathmandu as site of its secretariat.

13– Pakistan government arrests several opposition party leaders on eve of Independence Day and bans all public meetings; Benazir Bhutto arrested (14), sparking 6 days of riots which leave 16 dead.

31– Bangladesh President Ershad resigns as Chief of Army Staff; joins ruling Jatiya Party (1 Sept.); elected Party Chairman and Presidential candidate (2 Sept.).

September

3 Japanese Liberal Democratic Party grants Nakasone an extra year as party leader, allowing him one more year as prime minister.

5 US airliner with 389 passengers hijacked by 4 terrorists at Karachi

airport, after 16 hours security forces end hijack in battle that kills 18 and injures over 100.

20 Tenth Asian Games begin in Seoul, Nakasone begins 2-day visit to South Korea to repair relations.

25 China and USSR announce border negotiations will resume in 1987, after 9-year break.

28 Chinese Communist Party adopts 'Guiding Principles for Building a Socialist Society with an advanced Culture and Ideology'.

28– Polish leader Jaruzelski begins 3-day visit to China; signs cultural and scientific agreement (31).

October

2– Sikhs fail to assassinate Indian Prime Minister Rajiv Gandhi; Punjab police chief survives assassination attempt (3).

12 Britain's Queen Elizabeth begins 5-day visit to China.

15 Central Standing Committee of Taiwan's Nationalist Party approves reform plans to end martial law and allow formation of new political parties.

15 President Ershad wins Bangladesh presidential election, opposition parties claim widespread ballot-rigging and fraud.

15 USSR begins withdrawal of 6 regiments from Afghanistan.

21– East German President Honecker begins 6-day visit to China, re-establishing links between the countries' Communist Parties after a break of 20 years; signs economic and scientific co-operation accords (24).

22 US and 16 Pacific island nations agree on tuna-fishing treaty which also provides $10 million p.a. economic aid.

23 Marshall Islands adopt 'Compact of Free Association' with US, giving semi-independence and $30 million p.a. aid, US retaining authority over military activities.

28– Over 1,200 South Korean students arrested in 5-day riots at universities in Seoul and Pusan calling for President Chun's resignation.

30 Laotian Prince Souphanouvong resigns as President.

31 53 killed in riots in Karachi and Hyderabad, Pakistan.

November

3 US proclaims end of US Trust Territory of the Pacific.

3– Militant Tamil groups reject Sri Lanka government's June peace proposals; Indian Prime Minister Gandhi endorses the proposals (16); main Tamil leaders reject proposals but agree to attend negotiations (19).

5 3 US warships arrive in Qingdao to make the first port visit to China since 1949.

8– Japanese premier Nakasone begins 2-day visit to China; offers to help improve Chinese–South Korean relations (9).

10 Martial law in Bangladesh lifted after parliament grants President and government immunity from prosecution for acts under martial law.

17– South Korea reports assassination of North Korean President Kim Il Sung; North Korea denies report, and Kim appears at welcoming ceremony for Mongolian delegation (18).

20– Karmal resigns as President of Afghanistan's Revolutionary Council; replaced by Party General Secretary Najibullah (23 December).

23 Filipino President Aquino sacks Defence Minister Enrile, replaces him with his deputy, Rafael Ileto.

25– Soviet General Secretary Gorbachev begins 4-day visit to India; addressing joint session of Indian Parliament, makes proposals to reduce tension in Indian Ocean area (27); expresses hope for Afghanistan settlement in near future (28).

27– Filipino government and Communist rebel negotiators sign 60-day cease-fire agreement; cease-fire begins (10 Dec.).

December

2 UN General Assembly places French Pacific territory of New Caledonia on its Decolonialization List.

6 Ruling Nationalist Party wins Taiwanese elections by landslide in both Legislative and National assemblies.

9– India confers statehood on Arunachal Pradesh; China objects (11).

14– Police and army drug raid in Karachi, Pakistan, sparks 5-day riots which kill at least 180; premier Junejo's cabinet resigns (20) but is reconstituted with the same members (22).

15 USSR becomes first nuclear power to sign Pacific nuclear-free-zone (Rarotonga) treaty.

15– At Vietnamese Communist Party Congress, General Secretary Truong Chinh criticizes Party for country's economic problems; Chinh resigns (17); replaced by Nguyen Van Linh (18).

20– In China, student demonstrations calling for greater democracy spread from regional cities to Shanghai, and Beijing (23); government bans rallies in certain areas of Beijing (26).

23 New Zealand announces 3-year phased withdrawal of its troops from Singapore.

30 Special meeting of Japanese cabinet approves FY 1987 budget which increases defence spending beyond the ceiling (1% of GNP) set in 1976.

LATIN AMERICA AND THE CARIBBEAN

January

1– Haitian President Jean-Claude Duvalier sacks 4 cabinet ministers, including Chief of Police, in effort to calm anti-government protests; abolishes secret police (27); imposes a state of siege (31).

1 Aruba leaves Netherlands Antilles federation and becomes self-governed, Netherlands retains responsibility for defence and foreign affairs.

3 Mexico's President de la Madrid talks with US President Reagan in Washington, obtains US help for Mexico's debt crisis but differences on Central American situation remain.

11– Contadora group meets for talks in Venezuela; issues Caraballeda Declaration calling for US–Nicaragua negotiations (13); foreign ministers of Central American countries express support (15).

14 Vinicio Cerezo sworn in as President of Guatemala.

23 Bolivian workers stage 24-hour general strike to protest at government economic policy, President Estenssoro shuffles cabinet.

27 José Azcona sworn in as President of Honduras, pledges support for Contadora peace process.

February

1 Honduran armed forces chief Gen. Walter López resigns, replaced by Col. Humberto Regalado.
2 Centre-left candidate Oscar Arias elected President of Costa Rica.
7 Peruvian President García declares state of emergency in Lima and Callao to curb guerrilla attacks.
7 President Duvalier and family flee Haiti for asylum in France, 6-member military council under Chief-of-Staff Gen. Henri Namphy takes power.
7 Closing 4-day Cuban Communist Party Congress, President Castro announces 10 of 24 Politburo members and one-third of Central Committee replaced.
25 President Reagan requests Congressional approval for $100 million in military and humanitarian aid for Contras.
28 After 2-day talks in Uruguay, 8 Latin American foreign ministers agree to form international force to patrol and keep the peace on Costa Rica–Nicaragua border.
28 President Sarney of Brazil announces harsh economic measures, including introduction of new currency and price freeze, to control raging inflation.

March

7 Philip Habib appointed US Special Envoy to Central America.
7– In Ecuador, armed forces and air force chief Gen. Frank Vargas, accused of insubordination and asked to resign, refuses and takes over an air base; surrenders when his demand for resignations of defence minister and army chief is met (12); seizing another air base in Quito, demands overthrow of President Cordero (13); national state of emergency declared, troops storm base and arrest Vargas (14).
13 Deputy foreign ministers of Costa Rica, Nicaragua and Contadora countries, meeting in Costa Rica, agree composition of international border commission.
21 Chilean opposition groups stage day of protest against President Pinochet, violence erupts in Santiago and Valparaiso.
21 After several days of violent demonstrations Gen. Namphy, head of Haiti's ruling council, announces resignation of 3 council members linked with ex-President Duvalier.
22– Nicaraguan troops move into Honduras to attack Contra rebel camp; Reagan orders $20 million of emergency US military aid for Honduras (25); US Senate votes 53–47 to send $100 million to Contras, returns bill to House of Representatives, which had earlier rejected it (27).

April

7 Panama meeting of 13 Latin American foreign ministers fails to break deadlock in negotiations to sign Central American peace treaty before 6 June deadline.
27 In Chile, representatives of trade unions, academics and professional organizations form National Civil Assembly, call for restoration of democracy and social and economic reform.

May

16 Members of the Argentinian ruling Junta at time of Falklands war convicted of negligence and jailed (Gen. Galtieri for 12 years, Admiral Anaya for 14).

16– Presidential elections held in Dominican Republic; former President Joaquín Balaguer declared winner after delays caused by allegations of fraud (28).

16– Leader of rebel Democratic Revolutionary Alliance, Eden Pastora, leaves Nicaragua, applies for asylum in Costa Rica; asylum granted (3 June).

24– Five Central American leaders meet for 2 days in Esquipulas, Guatemala, agree to establish Central American Parliament.

25 Liberal Party candidate Virgilio Barco wins 63% of votes in Colombian presidential election.

June

1– Salvadorean President Duarte proposes resumption of peace talks; guerrillas accept offer (3).

1 Mid-term congressional elections in Ecuador bring swing to left-wing opposition parties, so conservative President Cordero loses support of Congress from August.

6– Contadora deadline for signing Central American peace treaty is not met; after talks in Panama new plan is proposed (7); rejected by El Salvador and Costa Rica for insufficient means of verification (12).

25 Brazil and Cuba re-establish diplomatic relations after 22-year break.

25– US House of Representatives approves (221–209) $100 million aid for Contras; International Court of Justice rules US aid to Contras violates international law (27).

July

2– 2-day general strike in Chile, called by National Civil Assembly, is widely supported, demonstrations in major cities broken up by police; 17 of NCA's 22-member council charged with violating state security (3); Pinochet announces he will stay head of state till 1997.

22 Mexico signs $US 10-billion rescue package with IMF.

29 Nicaragua files complaint at World Court against Honduras and Costa Rica for providing sanctuary for Contras.

29 Brazil and Argentina sign a series of economic accords as first step to establishing regional common market.

August

7 Virgilio Barco sworn in as Colombian President, calls for co-operation among Latin American nations to solve region's debt and poverty problems.

14 Uruguay and Brazil sign six economic accords at a ceremony in Brasilia, in step towards setting up South American common market.

19 Guatemala and Britain resume consular relations, broken in 1981, and intend to restore full diplomatic relations, broken in 1963, by end of year.

September

7– Failed attempt to assassinate President Pinochet of Chile kills 6 bodyguards; state of siege proclaimed, opposition leaders and journalists rounded up (8–).

9 Draft constitution presented to Nicaraguan National Assembly, half opposition deputies boycott debate.

14– Preparatory meeting for peace talks between Salvadorean rebels and government collapses when government refuses to remove troops from the town chosen for the talks; President Duarte attends on the appointed date but rebel leaders do not (19).

17 Mexico agrees rescheduling of $US 1.8-billion debt with 'Paris Club' of creditors.

October

5– Eugene Hasenfus (arrested survivor from weapons-carrying aircraft shot down by Nicaraguan troops) claims to be working for CIA; convicted by Nicaragua of terrorism and violating its security and jailed for 30 years (15 November); pardoned and returned to US (17 December).

27 UN General Assembly resolution calling for demilitarization of South Atlantic supported by 124 nations, US alone opposing.

29 Britain announces 150-mile fishery protection zone around Falkland Islands.

November

2– Foreign ministers of Brazil, Uruguay and Argentina discuss Britain's unilateral fisheries exclusion zone around Falklands; Organization of American States unanimously expresses concern at increased tension over Falklands, urges Britain and Argentina to resume talks on sovereignty (11); Argentina offers formal end to hostilities over Falklands and proposes global negotiations, Britain still refuses talks which include sovereignty (17).

15– Brazilian congressional and governorship elections bring big gains for government parties and endorse President Sarney's policies; tough economic measures introduced (21).

19 Nicaraguan National Assembly approves draft constitution.

December

1– Suriname imposes state of emergency over large areas of the country after rebel successes in south and east; government regains control of strategic towns in east (16).

6 Nicaraguan troops attack Contra bases in Honduras, Honduras asks US help to transport its forces to the area.

15 Ruling PNM party defeated in Trinidad and Tobago election, National Alliance for Reconstruction wins 33 of 36 seats and its leader, Arthur Robinson, becomes prime minister.

29 Guatemala and Britain resume full diplomatic relations.

EAST–WEST ARMS CONTROL

January

15 General Secretary Gorbachev proposes 3-stage plan to remove all nuclear weapons from earth by year 2000, extends Soviet unilateral nuclear test moratorium for 3 months.

February

20 At 40-nation UN Disarmament Conference Gorbachev indicates USSR is ready to accept on-site inspection to verify a nuclear test ban.

20 In Vienna Warsaw Pact tables response to NATO's December 1985 MBFR proposal.

23– President Reagan sends letter to Gorbachev welcoming his 15 Jan. proposal to remove all INF from Europe, but insisting that Soviet missiles in Asia also be included; US position tabled at INF talks in Geneva (24).

March

10 British premier Thatcher formally rejects Gorbachev's call to freeze modernization of Britain's nuclear force in the first stage of his 15 Jan. plan.

13 USSR announces it will extend its nuclear test moratorium beyond 31 March, until US carries out another test.

14– US proposes new verification measures to lead to US ratification of the Threshold Test Ban and Peaceful Nuclear Explosions Treaties, invites Soviet scientists to observe April US test; USSR rejects both offers (15).

29– Gorbachev offers to meet Reagan in Europe for summit to negotiate nuclear test ban; US declines (30).

April

11 USSR announces end of nuclear test moratorium after US test (10).

18 Speaking to East German Communist Party Congress in Berlin, Gorbachev proposes widening scope of conventional force reduction efforts to include all Europe from 'the Atlantic to the Urals'.

May

14 In televised speech on Chernobyl Gorbachev extends the Soviet unilateral nuclear testing moratorium to 6 August.

27 Reagan announces US will abrogate SALT II Treaty sub-limits in the autumn unless USSR corrects alleged violations.

29– NATO Foreign Ministers begin two-day meeting in Halifax, Nova Scotia; mandate High Level Task Force on conventional arms control (30).

June

11 Warsaw Pact Political Consultative Committee meeting in Budapest elaborates on Gorbachev's 18 April proposal, makes 'Budapest call' for conventional arms reductions from 'the Atlantic to the Urals'.

11 In START talks in Geneva USSR offers 30% cut in strategic offensive nuclear weapons in return for US commitment to remain in strict accordance with ABM Treaty for 15–20 years and confine space-based defence research and testing to the laboratory.

19 US House of Representatives urges President Reagan (256–145) to remain within SALT II limits.

30 NATO nations table new proposals at CDE in Stockholm.

July

15 British delegation at 40-nation UN Disarmament Conference in Geneva tables new proposals on chemical weapons ban verification.

At CDE in Stockholm Warsaw Pact nations drop demand for military activity in the air to be included in negotiations.

– US and USSR discuss SALT II at special session of Standing Consultative Commission; session ends abruptly in disagreement (30).

Reagan sends letter to Gorbachev proposing 50% reduction in strategic offensive weapons in exchange for US commitment to remain in ABM Treaty for at least 7½ years, further compliance conditional on elimination of all offensive ballistic missiles.

;– US and USSR begin 7-day meeting in Geneva on nuclear test ban.

ugust

1– US and Soviet arms-control experts begin official two-day meeting in Moscow to discuss current Geneva proposals.

8 USSR extends unilateral nuclear test moratorium to 1 January 1987.

9– Final session of CDE opens, USSR says it will allow on-site inspection of military activities.

eptember

3 US and USSR begin second round of talks on nuclear testing in Geneva.

5– US and Soviet arms-control experts begin second official two-day meeting on current Geneva proposals in Washington.

8 Sixth round of US–USSR Geneva talks begins, US tables new INF proposal.

2 After last-minute concessions by East and West, CDE agreement on reducing risk of war in Europe reached by 35 nations.

October

10 Reagan announces he will submit Threshold Test Ban and Peaceful Nuclear Explosions Treaties to Senate for ratification.

11– Reagan and Gorbachev begin two-day meeting in Reykjavik, Iceland; make substantial progress on INF and strategic offensive weapons reductions but meeting breaks up over disagreement on space-based defences (12).

November

4 Third CSCE Review Conference opens in Vienna, US Secretary of State Shultz and Soviet Foreign Minister Shevardnadze meet but fail to break Reykjavik deadlock.

25 Third round of US-Soviet nuclear test ban talks end in Geneva.

28 US breaks SALT II Treaty sub-limit by deploying 131st cruise-missile-armed B-52 bomber.

December

11 In 'Brussels Declaration on Conventional Arms Control' NATO foreign ministers agree in principle to new talks on reducing conventional forces in Europe from 'the Atlantic to the Urals'.

18 USSR announces it will end unilateral nuclear weapons test moratorium with first US 1987 test.